Applications of Particle Technology for Pharmaceuticals 2018

Topical at the 8th World Congress on Particle Technology

Orlando, Florida, USA
22 - 26 April 2018

ISBN: 978-1-5108-6968-4

Printed from e-media with permission by:

Curran Associates, Inc.
57 Morehouse Lane
Red Hook, NY 12571

CURRAN ASSOCIATES INC. proceedings.com

Some format issues inherent in the e-media version may also appear in this print version.

Printed by Curran Associates, Inc. (2018)

For permission requests, please contact AIChE
at the address below.

AIChE
120 Wall Street, FL 23
New York, NY 10005-4020

Phone: (800) 242-4363
Fax: (203) 775-5177

www.aiche.org

Additional copies of this publication are available from:

Curran Associates, Inc.
57 Morehouse Lane
Red Hook, NY 12571 USA
Phone: 845-758-0400
Fax: 845-758-2633
Email: curran@proceedings.com
Web: www.proceedings.com

TABLE OF CONTENTS

Some abstracts in this book refer to images that were not made available by the publisher

508322 Validation of CFD-DEM Simulation of a Continuous Tablet Coating Process

Abstract:

Tablet coating is a widely used unit operation in the pharmaceutical industry. Traditionally, large-scale drum coater systems are loaded with hundreds of kilos of tablets and rotate at relatively low rotation rates (Fr<<1). This leads to several hours of batch processing in order to obtain sufficient coating uniformity. The ConsiGma™ coater from GEA is a tablet coater which can be integrated into the GEA ConsiGma™ continuous process line. The ConsiGma™ Coater operates with low drum loads (3-7 kg) but very high rotation rates (Fr~1) leading to much shorter cycle times in the range of a few minutes. To achieve better coating results the tablets are jetted away from the rotating drum wall by so-called air knives in order to form a stable, gravity-free cascade. This air flow needs to be considered in a simulation model to accurately predict the coating process in the ConsiGma™ coater. Depending on the desired throughput rates there are three interchangeable sizes of the GEA ConsiGma™ Coaters.Interchangable means that the drum wheel of the next smaller drum will also fit and operate in the next larger housing (80/160mm or 160/320mm).

In this work we present a model to simulate the coating process in the ConsiGma™ 160 mm and 320 mm coater using CFD-DEM. The eXtended Particle System (XPS) is used for the DEM simulation and is coupled to AVL Fire® for the CFD[1]. The validity of the simulation results, as well as our method for modeling the non-sphericity of the tablets, are shown. The simulations are compared to experimental results, including the quantitative comparisons of the tablet flow and bed shape, as well as coating-related parameters, such as average coating mass on the tablets, standard deviation and coefficient of variation.

This work was funded through the Austrian COMET Program by the Austrian Federal Ministry of Transport, Innovation and Technology (BMVIT), the Austrian Federal Ministry of Economy, Family and Youth (BMWFJ) and by the State of Styria (Styrian Funding Agency SFG).

References

1. Jajcevic D, Siegmann E, Radeke C, Khinast JG. Large-scale CFDâ€"DEM simulations of fluidized granular systems. *Chem Eng Sci*. 2013;98:298-310. doi:10.1016/j.ces.2013.05.014.

507281 Modelling of High Shear Wet Granulation Using a Population Balance Framework

Abstract:

Wet granulation processes are difficult to reproduce and thus are often operated inefficiently which leads to high recycle rates in a continuous processes and high rejection rates in batch processes. A modelling approach can help to improve the efficiency of wet granulation processes significantly by understanding the effects of process parameters and material properties better.

In this study, a compartment-based model is used to model the wet granulation process, employing a one-dimensional population balance framework. For this modelling approach, the mixer is decomposed in three regions spray region, impeller region, and chopper region. Laboratory scale granulation experiments are conducted to estimate unknown parameters for empirical rate expressions. Finally, the model developed is validated using experimental data at different process scales. In order to assess the performance of the granulator model, the granulation end-point is predicted and compared to additional experimental results.

507217 Prediction of Pressure Filtration Performance in Systems with Pharmaceutical High Aspect Ratio Crystals

Abstract:

Filtration is a key product recovery process in the pharmaceutical industry as it facilitates the separation of suspended solid particles, such as Active Pharmaceutical Ingredients (APIs) from a liquid. Filtration is an intermediate processing step between crystallisation and drying, and it typically involves several washing cycles aiding the removal of impurities and the manipulation of the mother liquor solubility [1]. Pressure filtration is the most common separation technique and its performance is strongly influenced by the cake structure and formation, which in turn are highly dependent on particle properties such as shape and size distribution. Cake filtration in systems with wide particle size distributions (PSDs) and asymmetrical high aspect ratio particles is frequently quite time consuming at scale. Rising the system's pressure drop can accelerate the process, but it may also be counterbalanced by an increase in cake compressibility and/or particle breakage, both leading to denser packings [2]. The aim of this work is to study the cake filterability considering systems with high aspect ratio particles and various PSDs.

Lab scale filtrations have been conducted using a 200 ml, 35 mm diameter pocket filter using nitrogen pressure. Experiments have been conducted using dicalcium phosphate and mannitol. These materials were chosen because as pharmaceutical excipient the materials are commercially available with grades of different particle distributions. Filtration rates were compared for beds with different PSDs, at different applied pressures and for a settled bed or from a slurry. This approach was repeated using an API with high aspect ratio crystals that was wet milled to create three distinct PSDs and the filtration behaviour characterised in the same manner.

Modelling the filter cake structure is of crucial importance, since it allows for the accurate determination of cake properties. DigiDEM, a commercial Discrete Element Method (DEM) software, specially designed to deal with non-spherical shapes, with Lattice Boltzmann (LBM) built-in for drag force calculations [3], is used in this study to model the cake structure and consequently, evaluate the cake voidage as function of PSD and particle shape. DigiDEM is also used for the estimation of the particles' terminal velocities due to gravitational sedimentation. With validated cake voidage and terminal velocities predictions at hand, the improved Ruth equation (an expansion of Darcy's law) [4] is used in conjunction with lab-scale filtration data for the estimation of cake properties such as medium resistance, specific cake resistance and compressibility index. The estimated cake properties are then employed within the filtration model to predict the system response at higher pressures. Such a predictive model can be ultimately used for the successful system scale-up and control, and bring us a step closer to industrial adoption of a new Digital Design paradigm for pharmaceutical product and process design.

The authors gratefully acknowledge the support of the Advanced Manufacturing Supply Chain Initiative through the funding of the 'Advanced Digital Design of Pharmaceutical Therapeutics' project (Grant No. 14060).

References

1. Rushton, A., A.S. Ward, and R.G. Holdich, ***Solid-liquid filtration and separation technology.*** 2nd ed. 2000: WILEY-VCH Verlag GmbH.
2. Bourcier, D., J.P. Féraud, D. Colson, K. Mandrick, D. Ode, E. Brackx, and F. Puel, ***Influence of particle size and shape properties on cake resistance and compressibility during pressure filtration***. Chemical Engineering Science, 2016. **144**: 176-187.
3. Structure Vision Ltd 2015, viewed 11 July 2017, http://www.structurevision.com/.
4. Ruth, B.F., G.H. Montillon, and R.E. Montonna, ***Studies in filtration – I. Critical analysis of filtration theory***. Industrial & Engineering Chemistry, 1933. **25**: 76-82.

508074 Development of Continuous Manufacturing of Pharmaceuticals and Its Monitoring and Control

Abstract:

A novel continuous manufacturing system of pharmaceuticals with its monitoring and control systems was developed. The system consists of several unit operations such as mixing, twin screw granulation and spiral drying system. Performance of each unit operations as a function of each operating parameters was investigated experimentally. Also, a DEM (discrete element method) computer simulation was used to analyze the mechanism of twin screw granulation and a CFD (computational fluid dynamics) was used to analyze the mechanism of spiral drying system. Moisture content of granules in the twin screw granulation and the spiral drying was continuously measured and uniformity of API (Active Pharmaceutical Ingredient) content was also continuously monitored with a developed NIR (Near Infrared) sensor. Performance of the automatic manufacturing with feedback control systems of each unit operations was also investigated.

511465 UV Light-Inhibited Release for Time Controlled Growth Factor Delivery from TiO_2 Nanoparticles in a Molecular Hydrogel (Invited)

Abstract:

Tissue engineering scaffold materials mimic the natural extracellular matrix (ECM) to promote tissue regeneration and degrade naturally as healthy tissue grows. An ideal growth factor delivery system for regenerative medicine should operate synergistically with these tissue engineering materials. Spatial control of growth factor delivery is achieved by loading them into an implanted or injected scaffold, which acts as a reservoir for diffusional delivery to the surrounding tissue. The goal is often to provide a consistent therapeutic concentration through sustained, ongoing delivery, although depending on the application, many different delivery profiles can be required, such as rapid burst or pulsatile release. Advancement of the field depends on building a library of control techniques so as to be able to engineer any possible combination of different release profiles of multiple drugs or growth factors into a single material.

We have previously used chemical cross-linking via SMCC to provide a permanent ongoing presentation of growth factors by immobilising the growth factors to electrospun nanofibres in tissue engineering scaffold materials[1-3]. We have also used this SMCC attachment to temporally control their non-permanent delivery, by attaching polysaccharide chains to the growth factor to delay their release from self-assembling peptide (SAP) hydrogel tissue engineering materials[4].

Here, we have used SMCC attachment method and nanoparticle delivery to explore two new mechanisms for temporal control of delivery: attachment to a carrier nanoparticle embedded within a SAP hydrogel, and stimuli-responsive control of delivery from these nanoparticles. We report on a system in which nanoparticle bound BDNF shows greater release from the SAP hydrogel before UV treatment, at which point the unbound growth factor adsorbs to the SAP nanofibres. We also use this relationship and the UV-sensitive nanoparticles to induce a consistent release profile, using the UV response to counter the natural non-zero order delivery profile.

REFERENCES:

1. Wang T-Y, Bruggeman KAF, Sheean RK, Turner BJ, Nisbet DR, Parish CL. Characterization of the stability and bio-functionality of the tethered proteins on bioengineered scaffolds: implications for stem cell biology and tissue repair. J Biol Chem. 2014;289:15044-15051.
2. Potas JR, Haque F, Maclean FL, Nisbet DR. Interleukin-10 conjugated electrospun polycaprolactone (PCL) nanofibre scaffolds for promoting alternatively activated (M2) macrophages around the peripheral nerve in vivo. J Immunol Methods. 2015;420:38-49.
3. Horne MK, Nisbet DR, Forsythe JS, Parish CL. Three-dimensional nanofibrous scaffolds incorporating immobilized BDNF promote proliferation and differentiation of cortical neural stem cells. Stem Cells Dev. 2010;19(6):843-852.
4. Bruggeman KF, Rodriguez AL, Parish CL, Williams RJ, Nisbet DR. Temporally controlled release of multiple growth factors from a self-assembling peptide hydrogel. Nanotechnology. 2016;27(38):385102.

508525 Investigating Functionalized Particle – Microbe Interactions on 'Touch' Surfaces

Abstract:

Abstract: Disinfection of surfaces play a major role in deterring contact based pathogen transfer, thereby reducing infection rates. Common disinfection approaches employ the use soap and water (highly effective but not strong enough) and antimicrobial actives which are often harsh to the substrate/surfaces, and ultimately lead to antimicrobial resistance development. The primary objective of this study is to investigate a 'non-kill' based alternate disinfection approach focused on enhanced removal of microbes from skin via particle - bacteria interaction and stripping. The hypothesis that functionalized particles capable of interacting strongly with bacteria can be used to strip bacteria and disinfect surfaces more efficiently, has been verified with a set of functionalized particles for removing *E. coli* from a polymeric substrate. Particle momentum, particle – bacteria interaction strength, and particle concentration determined the disinfection efficacy. It was determined that modified particles could reduce bacterial concentrations to below detection limits (>4 log reduction) without inducing bacterial kill.

509029 Green Synthesis of Fluorescent Nanomaterials for Optical Bioimaging and Beyond

Abstract:

Fluorescent nanoparticles have received intense scientific attention and offer promising applications in many fields from optoelectronic devices to cancer treatment. The development of robust mass-scale nanoparticle synthesis methods is one of key materials challenges in moving towards fluorescent nanoparticles for commercialization and advanced applications. In this talk, I will summarize our work on green and scalable synthesis of fluorescent nanoparticles for multifunctional applications, including biomass-derived carbon dots for sensing and bioimaging, subcritical water nanonization of organic compounds for diagnosis and therapy, preparation of lanthanide-doped inorganic nanoparticles by high gravity method for transparent, flexible and luminescent composite films. A brief overview of this exciting field, along with some challenges and opportunities, will also be presented.

515286 Superparamagnetic Nanoparticles for Triggered Drug Release from Alginate Hydrogels

Abstract:

While magnetic nanomaterials have already been used in clinics for contrast enhancement in magnetic resonance imaging (MRI) (Singh and Sahoo, 2014), there has been no clinical approval for a drug delivery system containing such nanoparticles, yet. Magnetic nanoparticles currently investigated for their possible application in biomedicine are predominantly different crystalline polymorphs of iron oxides. For small enough crystal sizes, iron oxides exhibit the so-called "superparamagnetic" behavior, a feature combining high magnetization with very low coercive forces. Such superparamagnetic nanoparticles show great potential in therapeutic applications due to their ability to transform the energy of an alternating magnetic field to thermal energy in what is often called magnetic fluid hyperthermia (Jordan *et al*, 1999). These particles dissipate thermal energy by magnetic relaxation through the Brownian and Neel mechanisms. Therefore, aqueous suspensions at relatively higher nanoparticle concentrations (g/L), the so-called "ferrofluids", also increase their temperature in the presence of an alternating magnetic field (Sotiriou *et al*, 2013).

In this work, a composite multi-scale structure consisting of the biopolymer alginate, functional nanoparticles and a model drug is fabricated and analyzed. We demonstrate the highly scalable and reproducible synthesis of uniformly SiO_2-coated superparamagnetic Fe_2O_3 nanoparticles (Teleki *et al*, 2009), and evaluate their suitability as stimuli-responsive nanofillers in a drug-loaded biopolymer alginate matrix. The superior colloidal stability of the SiO_2-coated Fe_2O_3 nanoparticles over their uncoated counterparts and their dispersibility in aqueous suspensions facilitates their incorporation in alginate hydrogel microbeads. We perform detailed physicochemical and magnetic characterization on the hybrid alginate hydrogel beads and evaluate their potential in magnetic fluid hyperthermia and enhanced biomolecule release in the presence of an external alternating magnetic fields. We examine the hyperthermia performance of such multiscale particle structures in the presence of alternating magnetic fields and compare the release of dextran (a model biomolecule) in the presence and absence of external stimuli. The enhanced triggered release of dextran in the presence of magnetic fields further highlights the potential of such superparamagnetic SiO_2-coated Fe_2O_3 nanoparticles as a functional transducer in such systems. The possibility to externally stimulate drug release will open up new possibilities in intelligent, on-demand drug administration (Teleki *et al*, 2016).

Jordan A., R. Scholz, P. Wust, H. Fahling, R. Felix, *J. Magn. Magn. Mater.*, **201**, 413-419 (1999).

Singh, A., S. K. Sahoo, *Drug Discov. Today*, **19**, 474-481 (2014).

Sotiriou G.A., M. A. Visbal-Onufrak, A. Teleki, E. J. Juan, A. M. Hirt, S. E. Pratsinis, C. Rinaldi, *Chem. Mater.*, **25**, 4603-4612 (2013).

Teleki A., M. Suter, P. R. Kidambi, O. Ergeneman, F. Krumeich, B. J. Nelson, S. E. Pratsinis, *Chem. Mater.*, **21**, 2094-2100 (2009).

Teleki A., F. L. Haufe, A. M. Hirt, S. E. Pratsinis, G. A. Sotiriou, *RSC Adv.* **6**, 21503-21510 (2016).

508475 Toward Simulation-Based Design of Pharmaceutical Processes

Abstract:

Particle and fluid-particle simulations have been widely used to model a range of processes across many industries. This presentation will review application of simulations to common pharmaceutical processes including material transport and storage, blending, granulation, comminution, compaction, coating, and die filling. In many cases, simulations enable detailed analysis and flow visualization which provide insights into the effect of material properties and operating conditions promoting new designs and innovations and reducing inefficiencies and cutting costs associated with additional experimentation and scale-up studies.

Additionally, this presentation will review some computational models and techniques common to simulations involving particles. Recent advances in fluid-particle and particle models have significantly expanded the range of pharmaceutical systems and processes that can be described. Some of these advances include the modeling of particle mixes with wide variation in size, particle roughness, particle friction and particle cohesivity due to van der Waals or liquid bridging forces. Other model extensions include descriptions for particle aspericity that can also allow for particle breakage and flexibility. All of these complexities considerably affect the processing behavior of the particles. For example, surface moisture results in particle agglomeration, as is well known, but the particle aspect ratio determines the size and packing (solids volume fraction) of the agglomerates that are formed and the force required to convey or blend the particle mixture. And, when the solid volume fraction approaches the maximum packing density, the particle friction coefficient can be a more dominant factor than the liquid bridge force in determining the solid-phase shear stress.

508976 Novel Technologies to Improve the Bioavailability, Content Uniformity and Manufacturing of Pharmaceuticals

Abstract:

Approximately 65% of all prescription drugs are manufactured as solid dosage forms, which includes tablets and capsules. For very potent drugs, the amount of Active Pharmaceutical Ingredient (API) in the solid dosage form can be as low as 0.1% by weight. This very low API loading poses one of the biggest challenges in pharmaceutical product development: the control of dose uniformity. Low API content variability in the blend are highly desired and strictly enforced by the U.S. Food and Drug Administration. Another challenge in product development is that about 40% of newly discovered drug compounds have poor solubility which subsequently affects their bioavailability. An important aspect of pharmaceutical process development is the final product cost. As pharmaceutical companies strive to develop more affordable drugs, any possible elimination of lengthy and expensive unit operations becomes commercially advantageous. One group of such unit operations is associated with the control of API attributes (size, size distribution, shape, bulk density, etc.). These unit operations can include crystallization and various milling steps. The need for control of API attributes is solely dictated by the drug product development and usually is associated with desired improvements in blend uniformity or release profile. Having a formulation process that can make these and other steps unnecessary can provide a large advantage to pharmaceutical companies. In this talk, we will discuss a manufacturing method for solid dosage forms using fluidized bed impregnation of APIs onto porous excipients (inactive filler powders) that can potentially address all of the above challenges in drug substance and product development.

Fluidized bed impregnation can be summarized as the combination of three distinct processes taking place simultaneously. These include fluidization of the porous carrier, spraying API solution within the bed, which penetrates the excipient due to capillary forces, and drying of the porous particles causing the API to be deposited within. In this talk, we will present results showing that this process can deliver powder with high blend uniformity and tablets/capsules with high content uniformity. In addition, we show that the final API loading is not limited by its solubility in the organic solvent. Our results also show that fluidized bed impregnation does not depend on the nature of the API but rather on the nature of the excipient used. The product is an excipient impregnated with API that has flow properties that depend on the excipient and not on the physical properties of the API. In addition, we show that the product that is produced can significantly increase the dissolution kinetics of poorly soluble APIs. Implementing fluidized bed impregnation in drug manufacturing could allow for significant cost savings due to elimination of several unit operations.

513757 Particle Technology Impact on Pharmaceutical Product Performance and Process Design: Past, Present and Future

Abstract:

At the center of pharmaceutical product development lies the needs of patients. Each time a patient takes a drug, the patient expects consistent product quality and performance. These attributes are, in large part, determined at the product and process design stage. In recent years, particle technology has played a key role in advancing science-based approaches to product and process design shifting away from traditional empirical approaches. These science-based approaches drive efficiency and innovation in product development improving the pharmaceutical industryâ€™s ability to meet the patientâ€™s needs. Today particle technology applications span the drug development life cycle, from early stage materials characterization, to mid-stage product design and late stage process development and scale-up. This talk will present an industry perspective on how particle technology has transformed the product development paradigm. Real-world industrial examples will be presented. Current gaps and opportunities for future innovations will also be presented.

508397 Heterogeneous Liposome Assemblies: When Soft Meets Hard (Invited)

Abstract:

Liposomes, lipid bilayer vesicles, are almost routinely used as encapsulating agents for drug delivery. Individual liposomes typically serve as a protection layer that prevents the encapsulated drug from premature release and protects it from the environment. The release of the drug is then relatively slow and spontaneous. Although intensive research is carried on in order to prepare systems that are sensitive to changes in pH, temperature or external stimulus such as sonication or magnetic field, there is still lot of unexplored possibilities.

In our work we have designed liposome-based entities that consist from one or more liposomes as compartments and can contain also iron oxide nanoparticles (MgNPs) for remote control of the aggregate and/or silica nanoparticles to improve the loading capacity. We have tested several different approaches of forming the liposomal aggregates, including use of the alginate microcapsules as liposomes â€“ MgNPs vehicle, charged liposome aggregation due to high ionic strength of the buffer, using DNA amphiphiles as anchors between liposomes, and using polyelectrolytes such as poly-L-lysine as a glue for charged liposomes and MgNPs. This last approach was studied extensively and resulting aggregates, stabilized using the MgNPs, were tested for their ability to follow the magnetic field and to perform remotely triggered release. We have also tested the method of increasing the loading capacity of individual liposomes using porous silica nanoparticles (due to their good sorption properties). Such approach can significantly increase the loading capacity of liposomes and influence their response to external stimulus.

508570 Magnetic Iron Oxide Nanoparticles As a Local Source of Heat for Remotely Controlled Reaction

Abstract:

Magnetic iron oxide nanoparticles as a local source of heat for remotely controlled reaction

Ayse Beyza Aysan[1], Zdeněk Knejzlík[1,2], Aleš Zadražil[1], František Štěpánek[1]

[1]University of Chemistry and Technology, Faculty of Chemical Engineering Department, Prague, Technicka 5, 166 28 Prague 6, Tel. +420 220444320, E-mail: aysana@vscht.cz;

[2]Institute of Organic Chemistry and Biochemistry of the Czech Academy of Sciences of the Czech Republic, Flemingovo náměstí 542/2, 166 10, Prague 6

Nanomaterials have seen considerable attention in biochemical processes due to their physical, chemical and biomedical properties. One of the promising classes of nanomaterials are magnetic nanoparticles with the ability to undergo down-stream modification making them biocompatible, nontoxic, high level accumulated in the target region and acting as a local source of heating in the presence of an radiofrequency alternating magnetic field (RFAMF). High surface area and engineered functional groups make from the nanoparticles an ideal tool for magnetic resonance imaging (MRI), drug/gene carrier, tissue repair or heating mediators in hyperthermia.

The aim of the presentation is to use the colloidal magnetic nanoparticle suspension to initiate the polymerase chain reaction (PCR). Firstly, magnetic iron oxide nanoparticles (MNPs) were fabricated with relevant specifications and combined with PCR mixture containing all required components. Concordantly, we present a research about compatibility of MNPs with the PCR components and define the conditions for an effective PCR in the presence of high concentration of MNPs and regulate temperature cycles for PCR by RFAMF.

MNPs were fabricated successfully with different types of functional groups including sodium citrate dehydrate ($Na_3C_6H_5O_7.2H_2O$), 3-aminopropyl-triethoxysilane (APTES) and dextran 70 (from Leuconostoc spp., approximately Mw 70 kDa). After characterization of MNPs in terms of concentration, size distribution, surface morphology, and radiofrequency heating ability, they were used in PCR reaction to see the effect on the process. According to the results the response to the RF alternating magnetic field is proportional to the size and concentration of the nanoparticles. Dextran coated MNPs were stable in the PCR reaction mixture even in high concentration while the other types formed aggregates in excess of 1 μm. Experiments were focused on dextran coated MNPs in sufficiently high concentrations to achieve the required PCR temperature by exposure to an RF alternating magnetic field. Preliminary results showed that dextran coated MNPs in 5mg/ml concentration could increase the reaction temperature up to 97°C which was sufficient to start the PCR reaction under RF alternating magnetic field.

508617 Multi-Scale Engineering of Biomedical Made Nanomaterials and Devices By Scalable Flame Synthesis

Abstract:

In the last decades, nanostructured materials have demonstrated a considerable potential in many medical applications. This ranges from targeted drug delivery[1] to non-invasive medical diagnostics[2]. For example, the synthesis of stimuli-responsive nanocomposite able to more easily penetrate cell membranes and release specific molecules[1] is of great significance for delivery of anti-cancer drugs. Similar achievements have been obtained for medical devices[2, 3]. Portable diagnostic methods have been empowered by the utilization of tailored nanostructures[4]. For example, the latter can amplify relatively week biomarkers such as breath and body odour analysis resulting in novel non-invasive medical diagnostics opportunities[5].

We have recently shown that a large variety of nanoparticles such as photo-responsive TiO_2 and chemo-responsive SnO_2 can be rapidly synthesized by combustion of organometallic precursor solutions. This flame synthesis process is a very scalable technology currently utilized for the production of most nanoparticle commodities such as Carbon Black, fumed Silica, and Titania pigments. Its application to biomedical engineering has led to the synthesis of tailored materials with unique chemical and physical properties such as surface enhanced SnO_2-TiO_2 solid solutions[7]. These flame-made nanomaterials have demonstrated excellent gas sensing properties toward a large number of reducing and oxidizing gases[8]. This resulted in the first portable breath analysers for acetone detection, the main breath-marker for diabetes[2]. Optimization of this synthesis approach has enabled the fabrication of nanostructured surfaces for numerous applications including batteries, solar cells, and super-hydrophilic / hydrophobic[9] coatings[10]. In particular, such three-dimensional nanostructured surface allows enhancement of substrate - surrounding interaction and thus have potential to improve interface properties at sub-cellular level.

Here, we will present the synthesis of tailored nanostructured material by flame synthesis. We will focus on the application of this scalable approach for three main different applications, namely, the fabrication of stimuli-responsive nano-substrates for drug deliver and imaging, the assembly of biosensors for non-invasive medical diagnostics, and the rapid fabrication of biocompatible coating for medical implants. We will discuss the multi-scale optimization of the material nano-micro structure demonstrating that the interaction between organic and inorganic components can be optimized through suitable hierarchical configurations. References

1. H. Lee, et al., *Mol Pharm*, 2010, **7**, 1195-1208.

2. M. Righettoni, et al., *Anal Chim Acta*, 2012, **738**, 69-75.

3. M. Righettoni, A. Tricoli and S. E. Pratsinis, *Chem. Mat.*, 2010, **22**, 3152-3157.

4. A. Tricoli, M. Graf and S. E. Pratsinis, *Adv. Funct. Mat.*, 2008, **18**, 1969-1976.

5. G. Peng, et al. *Nature Nanotech.*, 2009, **4**, 669-673.

6. A. Tricoli, M. Righettoni and S. E. Pratsinis, *Nanotechnology*, 2009, **20**, 315502.

7. A. Tricoli and S. E. Pratsinis, *Nature Nanotech.*, 2010, **5**, 54-60.

8. A. Tricoli, et al., *Nanotechnology*, 2010, **21**, 465604.

9. A. Tricoli and T. D. Elmoe, *AIChE Journal*, 2012, **58**, 3578-3588.

509642 Targeting and Killing Melanoma Cancer Using Photoactivated Nanoparticles

Abstract:

Photodynamic therapy (PDT) uses photoactivated nanoparticles that are functionalized to target cancer cells and applies light to excite the nanoparticles which subsequently release reactive oxygen species (ROS) that set off a process resulting in cell-suicide of the cancer cells only. This study tested the polymer Poly[2,1,3-benzothiadiazole-4,7-diyl[4,4-bis(2-ethylhexyl)-4H-cyclopenta[2,1-b:3,4-b']dithiophene-2,6-diyl]] (PCPDTBT) for use as the photoactivated nanoparticle while also examining the Endothelin 3 ligand (ET3) for use as the functionalizing agent for targeting melanoma. This study was inconclusive. Although PCPDTBT was likely an effective photosensitizer for PDT, the results suggested that the functionalized nanoparticles did not target A375 melanoma cancer cells, likely due to the aggregation of the nanoparticles which exceeded the largest size that can be transported into cells at sizes of 100 nm and larger.

507737 A Compartmental Population Balance Modelling of Continuous Twin Screw Wet Granulation Using Mechanistic Process Kernels

Abstract:

Wet granulation is a critical process to create larger stable granules from fine powders in pharmaceutical manufacturing. Twin screw granulation (TSG) becomes increasingly popular as a method of wet granulation in the pharmaceutical industry as part of continuous manufacturing flowsheet due to its customized configuration, short residence time, relative insensitivity to changes in formulation properties and ability to produce porous granules suitable for downstream compression. TSG is very different from traditional batch high shear wet granulation (HSWG) and there is evidence that the balance of rate processes within is quite different. For example, breakage of large wet granules largely controlled by screw element geometry, and subsequent layering of unwet fines is a very important process in the TSG. Thus, rate expressions used in population balance models (PBM) for HSWG, or even for ball mills, may not be applicable to the TSG. This paper presents a compartmental population balance modelling of twin screw wet granulation with a mechanistic understanding of the rate processing kernels. To develop and validate the rate expressions, we use experiments explicitly designed to study a single rate process in a single type of screw element. As granule breakage is a key mechanism in controlling the granule size, breakage-isolated experiments are carried out in the conveying and distributive mixing elements. The dominant breakage mechanism in the conveying and distributive mixing elements is chipping and fragmentation respectively. Based on the breakage results in different elements, the breakage kernels including the selection function and the breakage function are mathematically formed. Other rate process kernels were also clearly specified in each compartment via mechanistic approach. Model results from the compartment based PBM are compared with experimental results for TSG of a model formulation over a range of liquid to solid ratios. Avenues to improve the predictive capacity by coupling PBM with DEM (Discrete element method) are also discussed.

508166 Understanding Powder Transfer in Force Feeders of Tableting Machines: A Comparison

Abstract:

Solid dosage forms, in particular tablets constitute the most frequent application form of active pharmaceutical ingredients (API) to patients. During tablet production, the three different processing stages, namely die filling, powder compaction, and tablet ejection, influence the tablet quality attributes to a different extent. The first stage, the die filling, is mainly defining the dose and its uniformity as it comprises the powder transfer from the blending hopper through the feeding system of the tableting machine into the dies of the rotating turret. For decades tablet press manufacturers have established a unique experience in pharmaceutical materials processing and tablet machine design however their "gut feeling" lacks a scientific basis. Although each of them have their own features and design components, only two main mechanisms for powder transfer from the hopper into the dies exist: gravity and force feeding. The latter mechanism is more popular and consists of one or more rotating paddle wheels in the feeder, which force the powder into the dies.

Supported by the increase of computational power, numerical methods that simulate the behavior of granular matter have gained increased attention also in the pharmaceutical industry within the last ten years. The *discrete element method* (DEM) is a numerical model that tracks the positions and velocities of individual particles through Newton's second law of motion thus providing a high resolution of particle based systems. Although some groups investigated the powder flow in force feeders through the DEM and improved process understanding, some limitations still exist. Those include the consideration of large particle sizes and material properties that are not resembling pharmaceutical powders and the neglect of the poly-disperse particle size nature of tableting blends. In this study, light is shed on the powder flow in force feeders of different tableting machines by means of the DEM. An iterative calibration procedure allowed the computation of a model direct compression formulation. First, the powder flow for two different sets of paddle wheel shapes in one tableting machine (FETTE) is studied. Afterwards the performance of this machine is compared with another one (KIKUSUI) to identify the setup that provides best product quality in terms of API content and tablet mass.

The computations were carried out using the open source DEM code known as LIGGGHTS. The material properties mimic a low dose strength (5% (w/w) API content) direct compression formulation including one API and six excipient particle sizes. The micro-mechanical particle properties were calibrated through an iterative process whereby the powder flowability of various commonly used direct compression diluents was compared with the numerical blend. Two different lab-scale tableting machines ("FETTE": FETTE 1200i, Fette compacting GmbH, Schwarzenbek, Germany) and ("KIKUSUI": KIKUSUI VIRGO, Kikusui Seisakusho Ltd. Kyoto, Japan) were compared. In the FETTE, the hopper is located in the middle of the feeder on top of the first paddle wheel, known as dosing wheel. The dosing wheel is transferring the powder to two paddle wheels below, namely the filling wheel which is the first filling station and which is rotating in opposite direction of the turret, and the reverse dosing wheel. The latter constitutes the second die filling station which is rotating in same direction as the turret with 24 punch stations. The reverse dosing wheel has angular shaped blades whereas for the dosing and filling wheel two different configurations are available. The first set includes rod like shaped blades that are alternating arranged and penetrating half of the powder bed height. The second set differs in two ways. First the blade shape is angular and second the blade is of the same height as the geometry demarcation. The KIKUSUI with 18 punch stations contains only two opposed rotating paddle wheels that are, in contrast to the

FETTE paddle wheels, overlapping. The feeding hopper is located above the left paddle wheel and the feeding plate contains an additional volume above the paddle wheels.

The numerical setup was adapted according to experiments which included an iterative filling of the system. Afterwards the actual high-speed tableting started by increasing the turret speed to obtain a throughput of 57,600 tablet per hour (40 rpm for FETTE and 53.3 rpm for KIKUSUI) at a paddle wheel speed of 30 rpm. The computations were carried out until in total 240 dies have been filled.

First of all, the powder flow in the force feeder of FETTE will be compared for the two different sets of paddle wheel shapes followed by a tablet quality comparison of the tableting machines FETTE and KIKUSUI.

The results show that the filling of the FETTE system itself caused particle size segregation for both paddle wheel sets that is then also found during steady state tableting. During steady state the following descending order of API content was found in different locations: for the rod like shaped paddle wheel combination: reverse dosing wheel (11.3%) > hopper (4.8%) dosing wheel (4.2%) > filling wheel (3.4%), for the angular shaped paddle wheel combination: reverse dosing wheel (14.2%) > dosing wheel (4.9%) > filling wheel (4.2%) > hopper (3.4%). These API content differences were also transferred into the die as follows.

In case of the rod like shaped paddle wheels, 85.3 mg of API mass entered the die underneath the filling followed by an additional 125.6 mg during reverse dosing wheel zone. Afterwards 5.3 mg API were scraped during dosing cam in which the lower punch was moved upwards by 3 mm to obtain a more confined powder bed that would subsequently be subjected to the compaction process. In contrast, for the angular shaped dosing and filling wheel combination more than double the API amount (175.6 mg) entered the die from filling and 120.8 mg from reverse dosing wheel zone. In other words, in the reverse dosing wheel, which was the same for both paddle wheel shape sets, a similar amount of API was transferred into the die. After scrapping of 16.8 mg API mass during dosing cam, 36.1% more API mass (279.6 mg vs 205.5 mg) compared to the rod like shaped paddle wheel set were in the filled dies. These differences between API masses can be explained by two phenomena. First a different API content in the paddle wheel zones (confer above) was found and the paddle wheel shape influenced the extent of the force feeding effect. As mentioned in the shape description, the angular shaped filling wheel was transporting the complete height of the powder bed during rotation in contrast to the rod like shaped wheel in which the blades only penetrated half of the powder bed height. Thereby more mass (angular shape) could be transported from the force feeder into the dies. As a consequence the blend mass flow rates were 32.8 g/s and 25.7 g/s for the angular and rod like shaped filling wheels, respectively. Although the final tablet masses were only of minor difference (4.3%), the API contents difference of 21.7% exceeded this value unproportionally, indicating that incorrect API dosing might constitute a quality risk. However further analysis is necessary to verify this observation in long term production (> 240 dies).

In case of the KIKUSUI tableting machine, particle size segregation during steady state tableting revealed significant different API contents throughout the feeder with 5.0%, 1.2%, and 4.3% in left, and right paddle wheel zones, and hopper, respectively. The first particles entered the die underneath the right paddle wheel zone resulting in 133.3 mg API mass. As a significantly higher API content was found in left paddle wheel zone, a 1.6-fold more API amount (216.2 mg) compared to right paddle wheel zone was filled in the second die filling half. The API particles entered the die later thereby the majority of them were located in the top portion of the die thus they were scraped to a considerably high extent during dosing cam (67.4 mg or 19.3%). At the end, a final API mass of 282.1 mg was obtained which was higher compared to both paddle wheel sets in FETTE.

The comparison of all tableting machines gave the following order in terms of tablet mass: KIKUSUI 5.0 +/- 0.1 g, FETTE angular 4.9 +/- 0.1 g, and rod like shaped paddle wheel set 4.6 +/- 0.2 g. With respect to API content 5.5 +/- 1.0 %, 5.6 +/- 0.8%, and 4.6 +/- 1.1% were obtained for KIKUSUI, FETTE angular, and rod like shaped paddle wheels, respectively. Hence most significant differences between the tableting machines could be revealed for API dosing.

These distinctive differences in tablet mass and API content showed that the powder transfer from the hopper into the dies was significantly influenced by the force feeder design, in terms of number of paddle wheels and their shape in addition to the housing geometry. This study shows that the tablet quality is not only affected by the formulation and process conditions (e.g. operating speeds) but the tableting machine exerts a major impact. In next studies the powder flow mechanism will be analyzed more thoroughly by particles' velocity and coloring, residence time, and size segregation comparison to provide a more detailed understanding.

508215 A Development of Combined PBM-RTD Technique to Simulate Continuous Wet Granulation Systems for Pharmaceutical Manufacturing Processes

Abstract:

The healthcare requirements of the developing countries in the world are expected to grow in keeping pace with their burgeoning populations. In efforts to meet the current and future projected supply-demand figures, there is a paradigm shift in pharmaceutical manufacturing towards producing the drug products (most commonly tablets) in a continuous operation as opposed to traditional batch methods of manufacturing in the industry. Granulation is a key unit operation that is responsible for the reduction in the segregation of the components, increase in the blend composition and, is responsible for producing particles with homogeneous particle size distributions. In an effort to predict the outlet compositions of the particles of a continuous granulator, the study of residence time distributions (RTDs) is crucial in order to predict the effects of variabilities in the inlet parameters such as feeder fluctuations and blend inhomogeneity on the outlet granule particle sizes and compositions. Population Balance Models (PBMs) are developed and extensively employed in modelling Wet Granulation systems in order to track the particle size distributions (PSDs) and porosities of the particles. Therefore, this work has initially studied two (2) different RTD models, compared them, and has applied them to a continuous granulation system. The RTD models are being combined with a PBM model in order to develop a model that quantifies the processes and attempts to describe the phenomena in as mechanistically as possible.

RTD models in granulation literature have been modelled in similar vein as the models for continuous liquid reactor systems [1]. The models account for plug flow volume fractions (conveying of material), number of continuous stirred tanks (back-mixing within mixing and kneading sections), and dead volume fraction (stagnant material within the unit operation). RTD models for hot-melt extrusion process (HMEs), which occur in similar equipment as granulation in polymeric media, have been given similar treatment in literature. The HME models too are described as reactor trains, and have even been modelled as cascading series of stirred tanks with recycle flows added from the subsequent reactor to the preceding tanks [4].

The present study has initially done a case study of two kinds of cascading reactor models for granulation operation. The first one, that has been published and studies for HME systems assumes that the ratio between the amount of material recycle to the previous tank to the amount of fresh feed coming into the current tank is constant for all the tanks in the cascade. This model holds true when the flow of material in either direction is uniform in any particular local mixing section, but the degree of mixing decreases gradually as the material is conveyed forward. The other approach that was developed in this work models the system such that the amount recycle to a previous stage is a fraction of the total inlet coming into the section. This accounts for the case when the degree of mixing is uniform in all the sections of the operation but the amount of material exchanged is different.

The two different models are being developed and studied in detail. They are applied to an experimental case of continuous granulation whereby, the model predictions and their closeness to experimental results are being compared.

An initial experimental study on a twin screw granulator has been performed using a blend of caffeine, micro-crystalline cellulose (MCC), lactose and polyvinylpyrrolidone (PVP) across a DOE consisting of variation in powder feed rate, liquid content and screw rotation speed. The RTD experiments were carried out on the DOE using the pulse method of tracer study using caffeine (one of the components of the formulation). The exit concentration was measured in real time using an inline near infrared probe (NIR). The NIR was calibrated with blends containing the aforementioned components in different combinations of proportions in order to measure real time changes in caffeine concentration due to addition of pulse tracer. The initial test results showed that the mean residence time (MRT) decreased with an increase of either the screw rotation speed (RPM) or increase in the feeder powder flow rate.

Further refined studies were carried out using the same components and experimental methods on the full DOE and the data is being fitted to the various RTD models. It is hoped that the RTD model would rely less on the similarity and approximation to the various TIS models developed for traditional liquid reactor systems and would be more inherently mechanistic and derived from basic principles applicable to granulation and pharmaceutical material systems.

In order to achieve the above mentioned goals, it is envisioned that the traditionally developed PBM equations [3] would be employed for estimating the particle size growth and porosity change as the primary powder particles are transformed into granules. The PBM equations take into account the binary aggregation of the smaller particles, breakage of larger particles and granules into smaller ones due to shear from the screws, liquid addition due to binder spraying, liquid evaporation due to hot air drying, and gas consolidation due to compression of the particles. The aggregation rate is generally considered to be proportional to the number of particles and the fractional area of the wetted region in the colliding particles. The breakage rate is primarily proportional to the screw speed and a power law factor to the particle size. These four elementary particulate rates are responsible for the changes in particle sizes and porosity.

Combining PBM with RTD models hasn't been reported yet in pharmaceutical granulation modelling literature. However, a combined approach has been employed in modelling the rate processes of colloidal aggregation of beta-lactose globulin in calcium chloride solutions [2]. Therefore, it is envisioned that a similar approach would lead to a combined PBM/ RTD model for simulating continuous pharmaceutical granulation. It is hoped that the model developed would be as mechanistic as possible that would lead to reduction in dependence on experimental data and parameter fitting/ assumptions often involved in the current semi- empirical models.

Such models would then be employed in the simulation of a continuous pharmaceutical process line with integrated process control models. The integrated models would then be useful for designing quality manufacturing setups which would provide the advantages of lower equipment footprint, reduced manufacturing cost and reduced human intervention in the entire pharmaceutical manufacturing process thereby, leading to high- quality pharmaceutical products at affordable prices to the uninsured and under-insured consumers in the developed and developing countries.

References:

[1] A. Kumar *et al.*, 'Conceptual framework for model-based analysis of residence time distribution in twin-screw granulation', *European Journal of Pharmaceutical Sciences*, vol. 71, pp. 25–34, Apr. 2015.

[2] N. Erabit, F. T. Ndoye, G. Alvarez, and D. Flick, 'Coupling population balance model and residence time distribution for pilot-scale modelling of β-lactoglobulin aggregation process', *Journal of Food Engineering*, vol. 177, pp. 31–41, May 2016.

[3] A. Chaudhury, 'Mechanistic modeling, simulation and optimization of wet granulation processes', Rutgers University - Graduate School - New Brunswick, 2015.

[4] J. Puaux, G. Bozga, and A. Ainser, 'Residence time distribution in a corotating twin-screw extruder', *Chemical Engineering Science*, vol. 55, no. 9, pp. 1641–1651, May 2000.

508511 Development and Application of Coupled CFD-DEM Models for Complex FLUID Particle Systems (FIBROUS MATERIAL, SLURRIES ETC...) in Industrial Scale Processes

Abstract:

Many processes in the chemical industry involve the both fluids and particle flow.. Slurry mills, cyclones, desanders, pneumatic conveyors, grains drying and sorting, fluidized beds and catalytic reactors are just a few examples of granular-fluid systems. In order to predict the behavior of such systems accurately using a first principle modeling tool, it is important to model the effect of fluid flow on the particles. Design and scale-up, as well as optimization of such processes, require a deep understanding of the thermo-hydrodynamics, which is determined by the particle-level interactions between the fluids, particles, and boundaries.
The complexity of the fluid-solid flow present in these systems makes modeling them a challenging task. The primary source of difficulty is the wide range of length scales existing in these system, as the largest structures are a few meters whereas the momentum, heat and mass transfers occur at the micrometer scale of the interactions between particles and fluid and particles. Although, there are some inherent challenges in developing CFD-DEM models, such as large computational run times, the advent of fast GPU hardware and advanced algorithms are mitigating these limitations.

In this work, details of the DEM-CFD integration between two commercial softwares, Rocky® (DEM) and ANSYS Fluent® (CFD) will be addressed. In particular, the coupling formulation for non-spherical particles shapes and a wide variety of application that includes flexible fibrous material, slurry flows etc... will be evaluated. In addition, a few application examples, including both one-way and two-way coupling cases will also be presented. These models capture the discrete nature of the particle phase while maintaining the computational tractability. This is accomplished by not solving the detailed fluid field at the particle level, which enlarges the range of equipment and processes that can be studied with numerical simulations.

507311 Investigation the Complex Gas-Solids Flow Characteristics in a Fluidized Bed with Wurster Tube By Process Tomography and CFD Simulation

Abstract:

Wurster type fluidized beds are widely used in pharmaceutical industry for pellets coating. They are considered to be the best choice for pellets coating as they can produce a superior film compared with other coating strategies. For a Wurster type fluidized bed, there exists a concentric cylindrical tube which is called Wurster tube inside a conical chamber, and the flow region is mainly divided into two parts, i.e. the coating zone inside the Wurster tube and annulus zone outside the Wurster tube. A dual-fluid nozzle is placed in the bottom of the Wurster tube for spraying the coating solutions. In this research, process tomography sensors including electrical capacitance tomography (ECT) and microwave tomography (MWT) are designed and used to investigate the complex gas-solids flow characteristics in a Wurster type fluidized bed for coating and drying. One set of 12-4-8 combined electrical capacitance tomography (ECT) sensor was designed and used to monitor the solid concentration inside the chamber. The ECT measurement was conducted synchronously both inside and outside of the Wurster tube, i.e. coating zone and drying zone, to achieve a complete monitor of fluidised bed process with Wurster tube. The MWT sensor has 16 antennas and the measurement frequency is in the range of 1.0 GHz and 2.5 GHz. A series of tests were carried out by varying the operational parameters including the Wurster tube gap, air flow velocity and particle mass load. Different flow patterns as well as flow stability were evaluated based on the tomography measurement results. To evaluate the flow characteristics inside the bed, fast flourier transform (FFT) was applied to investigate the fluctuation in the coating zone and drying zone. The hydrodynamics of gas-solids flow with different moisture content were revealed by the tomography images and spectrum structure based on FFT analysis.

To verify and optimize the measurement result, computational fluid dynamic (CFD) based on the kinetic theory of granular flow (KTGF) is used to investigate the flow characteristics in the same bed. Different drag models, i.e. the energy-minimization multi-scale (EMMS) model, Wen-Yu and Gidaspow model, are used to investigate the gas-particle interface effect on the flow characteristics. Key process parameters, including the solids concentration, solids velocity and pressure drop, are analyzed and verified by the measured results. Both the CFD simulation and tomography measurement results provide valuable information to reveal the hydrodynamic characteristics in the Wurster type fluidized bed. With the lab-scale fluidized bed, the results of Wen-Yu and Gidaspow drag models have a good agreement with the ECT measurement. The results also indicate that the design of the Wurster tube has significant effects on the flow. Based on the CFD simulation and ECT measurement results, optimized parameters are chosen to improve the solids circulation efficiency and coating quality.

507053 On Designing Functional Microparticles for Encapsulation and Controlled Release

Abstract:

The talk will provide an overview on some the research conducted at the Monash Advanced Particle Engineering Laboratory (http://www.mapelab.com/home.asp) and the Biotechnology and Food Engineering Group (BFE). The research group at Monash University utilizes a microfluidic drying approach to generate uniform microparticles with tightly controlled characteristics and sizes in a scalable, almost waste-free process. Microparticles of different materials with tailor-made properties and functionalities could be produced for encapsulation and controlled release. The effects of formulation (pH, concentration, dopants, solvents) and process parameters (droplet size, drying time) on microstructural properties and controlled release functionalities were investigated for silica-based and polymer-based microencapsulates. The ability to control the drug release kinetics of polymeric microparticles by manipulating their microstructures was demonstrated using polymeric aqueous dispersions Eudragit® RS and Eudragit® NE as matrix materials. The microstructures of particles and corresponding drug release mechanisms could be tuned by different blending strategies, such as the incorporation of highly hydrophilic ingredients (e.g. lactose) to accelerate release, or by using alginate or acid-hydrolysed tetraethyl orthosilicate (TEOS) as an additional barrier in the matrix. Microstructures with defined core-shell structures were spray dried using Eudragit® RS /silica system. The uniform core-shell microencapsulates with different shell thickness were assembled in a single step from a homogenous precursor, illustrating the ability to tailor particle structures and release rates of the active ingredient. Other examples including microencapsulation of curcumin, DHA, and astaxanthin using dairy-based proteins.

The ability to produce highly uniform particles with this unique approach has contributed to the understanding of how particles with different microstructures are formed, and how the physicochemical properties are related to their functionality. The underlying principles of particle formation mechanisms are applicable for other functional particles including mesoporous and magnetic particles. The strategy is useful for designing spray-dried particles of targeted properties in a scalable manner using multi-nozzles microfluidic dryer (to generate uniform particles), or with a conventional spray dryer.

508045 Additive Manufacturing of High Drug Loaded Extended Release Tablets via FDM 3D Printing

Abstract:

3D printing technology is an active research area in personalized medicine. Coupling of fused deposition modeling (FDM) based three-dimensional (3D) printing with hot melt extrusion (HME) has the distinct advantages of tailoring dosage forms, enhanced bioavailability of poorly water soluble drug as well as complex dosage geometries. However, there remain major challenges in forming 3D compatible filaments, due to the limitations such as low drug loading, challenges in developing required formulations suitable for HME along with inherent process limitations such as the need for high temperatures, limited options for pharmaceutically compatible polymers, inadequate filament mechanical properties such as brittleness, and achieving very good drug product content uniformity (CU). This work explores the ability to fabricate filaments that are compatible for FDM-3D printing with a wide range of drug loadings while retaining proper mechanical properties and good drug CU within the filaments. Griseofulvin (GF) and hydroxypropyl cellulose (HPC) was used as the model BCS Class II drug and the polymeric matrix, respectively. Dry GF and HPC powders were pre-mixed having up to 40 wt.% GF and dispersed into the molten polymeric matrix via HME. The produced filaments were used to produce cylindrical shape tablets in FDM 3D printing. All printed tablets showed good CU and sufficient mechanical properties up to 25 wt.% GF while high drug-loaded filaments i.e., 40 wt.% GF created processing difficulties in a 11 mm twin-screw extruder (TSE). TSE processing was successfully done for four different drug loads (5, 10, 15, and 25%) to examine their influence on mechanical properties as well as drug release. Increasing drug load from 5 to 25 wt.% in extruded filaments resulted in enhanced modulus of elasticity from 235.89 to 298.15 MPa, respectively. Using a FDM-3D printer, mini tablets were printed containing up to 25wt.% GF (15 mg). The printed tablets showed improved CU (RSD <2%), and acceptance values (AVs) for all printed tablets were less than 8. Drug loading in both filaments and 3D printed tablets showed similar values, indicating no degradation during extrusion, matching with DSC/TGA results. In vitro drug release testing for all 3D printed tablets revealed extended release without any burst effects for up to 12h. In summary, the filaments produced via HME provided high flexible dosing capability as well as required mechanical properties and could be used to prepare FDM 3D printed tablets that can be potentially used for personalized medicine.

506718 Spray Dried Submicron Sized Particles for Pharmaceutical Application

Abstract:

The low water solubility of newly identified active pharmaceutical ingredients necessitates new strategies, since the dissolution of the drug in an aqueous solution is essential to ensure pharmacological response. So far, different approaches and strategies were developed to increase the bioavailability (physical and chemical modifications) by increasing their aqueous solubility and dissolution rate. This work focusses on the generation of submicron sized particles (0.1-1 µm) of the pure drug. Due to the increased specific surface area of submicron sized drugs, the dissolution rate increases according to the Nernst Brunner equation. Particle size reduction down to the submicron range also leads to increased saturation concentration, described by the Ostwald-Freundlich equation.

One possibility for the generation of submicron sized particles is the spray drying process. The pure solution is firstly atomized, dried and finally separated from the drying gas, usually by a cyclone. Spray drying is a single step process, beginning with the liquid to the dry powder. Particle size, shape and morphology can be adjusted by controlling the spray drying conditions.

However, spray drying for the preparation of submicron particles is challenging. The particle size during spray drying is mainly limited by the droplet size and the feed concentration. Thus, small droplet sizes with the use of higher concentrations are preferable. Conventional atomizer show limitations in the generation of small droplets ($d_{50,3}$<3 µm). Next to the atomization step, the precipitation of particles down to the submicron range with conventional cyclone particle separator is limited.

In this work, the advantages of the spray drying process is used and additional methods were integrated to generate a robust process with a long-term stability for the preparation of submicron particles.

To overcome the challenge of the production of fine droplet, a two-fluid nozzle with internal mixing is combined with a cyclone droplet separator. Two-fluid-nozzles show advantages regarding atomization of higher viscous solutions and concentration, thus generating the smallest droplet sizes ($d_{50,3}$=10 µm), compared to other atomization techniques. According to a specific cut-off diameter of the cyclone droplet separator, bigger droplets were separated and returned back into the liquid feed. Thus, only droplets smaller the cut-off diameter reaches the following drying stage. The dried submicron sized particles are then separated from the drying gas. This is done by using an electrostatic precipitator. The particles were collected at the inner wall of the precipitator.

The experiments were figured out with a 10 w.-% mannitol (Roquette ® Pharma, Lestrem, France) solution as a physiologically uncritical model substance. First, the volumetric droplet size distribution of the conditioned aerosol after the cyclone separator process was analyzed by laser diffraction technique (Malvern Instruments, Spraytec, Malvern, Worcestershire, United Kingdom). The mean volumetric droplet size $d_{50,3}$ and the span value were extracted from the distribution. Different liquid-to-gas mass flow ratios in the range of 0.8 to 1.8 were used and the gas pressure was held constant at about Î"p=5 bar. It could be shown, that the mean volumetric droplet size is constant (1.9 Âµm) for different liquid-to-gas mass flow ratios Âµ. Thus, the droplet size only depends on the cut-off-size of the droplet separator. The droplet size distribution generated by a two fluid nozzle depends on the characteristics of the liquid feed, the liquid-to-gas mass flow ratio and the gas pressure Î”p. In addition, the mean volumetric droplet size was measured for different Polyvinylpyrrolidone K30 (BASF; Ludwigshafen, Germany) solutions with viscosities of 9, 20 and 41 mPas. Even for different viscosities, the mean droplet size depends only on the cut-off size of the separator, and is still constant at about 1.9 µm. With this concepts, fluctuation in the primary aerosol were compensated, thus a well defined spray is introduced into the drying chamber.

The mass flow of droplets in the conditioned aerosol is determined by the amount of droplets smaller than the cut-off size in the primary aerosol. Thus, the useful yield of the primary aerosol essentially determines the particle yield after drying.

A compact spray dryer was designed. The aerosol conditioning step with the two-fluid nozzle and the cyclone droplet separator was combined with the drying step and the electrostatic precipitator. First drying experiments with a 10 wt. - % mannitol solution confirm the generation of submicron particles with a mean volumetric diameter of $d_{50m,3}$=0.7 µm.

Spray drying experiments will be realized by using pure pharmaceutical model drugs. The particle size will be determined by laser diffraction and the morphology and shape will be analyzed by evaluating scanning electron microscopy images of the dried particles. In addition, the crystallinity of the particles will be determined by X-ray diffraction. Finally, in order to test the in vitro drug release of the spray dried submicron sized particles, dissolution tests will be presented.

506529 Efficient Precipitation of Spray Dried Submicron Particles for Pharmaceutical Applications Using a Two-Stage Electrostatic Precipitator

Abstract:

The bioavailability enhancement of poorly water soluble drugs can be achieved by the decrease of the particle size of active pharmaceutical ingredients (API). An innovative process is the spray drying with spray conditioning to produce submicron particles. A resulting challenge is the collection of these dispersed particles from a gas flow. The separation in electrostatic precipitators (ESP) is a common technique for air purification purposes where powder particles can be collected rather independently from particle size. An adapted ESP design is necessary to achieve high separation efficiencies and a robust process.

The ESP design in this work uses the precipitation method of Penney filters which separates charging and collection into two zones. Submicron particles tend to follow the gas flow in appearing turbulences. The setting of a laminar flow in the second zone prevents the appearance of eddies, provides the necessary time for the particles to move to the collection electrode and enhances the efficiency compared to single stage ESPs.

The dimensioning is depending on various assumptions which have to be revised. Its design of perpendicular mounted electro-polished stainless steel pipes allows an easy cleaning and simplifies the setup. The gas flow is induced by a fan which draws the air through the precipitator. Particularly with the use of active pharmaceuticals this formation prevents the API to leak before it gets separated.

Several experiments are conducted with the model substance mannitol (PEARLITOL® 160 C, Roquette Pharma®, Lestrem, France) to examine the ESP and characterize its behaviour in long term tests. The crucial parameters in the charging process are the residence time as well as the electric adjustment. These constraints were examined to enhance the product recovery. Based on these tests it is possible to determine a suitable charging length as well as the dimensions of the precipitation zone.

The efficiency of the ESP is measured with an optical particle counter (1.109, GRIMM AEROSOL Technik, Ainring, Germany). To ensure a characterization of the precipitated material the collected powder is studied in a scanning electron microscope (SEM, H-S4500 FEG, Hitachi High Technologies Europe, Krefeld, Germany).

The separation process is conducted with spray dried mannitol submicron particles at a voltage level of 20 kV and a current density of 7.5 mA/m^2 at start. The precipitation efficiency decreases after 3 hours due to the dust layer and the dust resistivity which is increasing. To enhance the precipitation time it is necessary to examine the first stage of the ESP. In theory the current flow is transferred mainly in the ionization stage. By cleaning this part of the filter during the separation process, the resistivity can be decreased and particles can be treated with a constant amount of charge. An increasing dust layer leads to a higher resistivity and therefore to a lower current flow. For further investigations a cleaning step of the charging zone was added to enhance the separation duration. The results show a separation efficiency in a continuous experiment of higher than 99% for 10 hours for 500 nm particles. These high efficiencies result in a production rate of 1 g/h. In conclusion the first stage is the limitation in the separation process. If a particle doesnâ€™t receive charge in the first stage it cannot be deflected in the second stage. In an upcoming work the functionalization of the separated particles shall be examined.

509616 The Compression and Compaction Behaviour of Pharmaceutical Powders and Their Binary Mixtures

Abstract:

Powder compaction is a very important production process in the pharmaceutical industry due to the prevalence of tablets as solid dosage form. Therefore, various scientific investigations focused on the investigation of the powder compaction process in the past decades. Nevertheless, the physical processes at particle level during powder compression are still not fully understood and this is the reason for the mainly empirical formulation and process development. The missing process understanding can be attributed to the complexity of the powder compaction process due to the high number of influencing parameters on the one hand and different acting micro-processes on the other hand. Exemplary influencing parameters are particle size distribution, particle shape and deformation behaviour of the raw materials as well as compaction stress and velocity [1; 2]. All these parameters affect the contact area and the number of contacts between the particles and hence, the mechanical and structural properties of the tablets.

Furthermore, it is well known that different micro-processes occur during powder compaction [3–6]. At the beginning of the powder compaction and thus, at low compression stresses, particle rearrangement is the main deformation mechanism and causes the reduction of inter-particulate pores due to the filling of large pores. The increase of compaction stress leads to the elastic and plastic deformation as well as fragmentation of single particles, which, in addition influences the contact area and the number of contacts between the particles. Additionally, elastic deformation of the molecule lattice occurs and causes the decrease of solids volume and thus, an increase of solids density with rising compression stress [7–9]. This phenomenon is referred to as solids compressibility. The different deformation mechanisms do not appear successively, but appear in parallel. The proportions of the single mechanisms differ in dependence on the used materials and process parameters and a differentiation between the mechanisms during the process is difficult or rather impossible. At the decompression phase, elastic recovery of the particles takes place and can lead to the destruction of weak inter-particulate bonds and to the occurrence of capping and lamination phenomena [6; 10; 11].

The missing physical process understanding requires the comprehensive characterization of the compressibility and compactibility of raw materials and the derivation of characteristic material parameters enables a more valid prediction of formulation and process development. The compressibility of a powder can be described by the relation between volume or porosity decrease and compaction stress [4]. Instrumented tablet presses enable the determination of this relationship directly during the compaction process (in-die). Various mathematical models for the description of this relation were developed, such as the models of Heckel, of Kawakita and of Cooper and Eaton [12–14]. Nonetheless, these models are mostly empirical and often describe only one part of the compaction process. Additionally, the existing mathematical models do not consider the phenomenon of solids compressibility.

In this work, the compressibility and compactibility of pharmaceutical powders with different deformation behaviour were investigated using in and out-die analysis. The compaction experiments were performed using the compaction simulator Styl'One Evolution (Medel'Pharm, France), which is a single station tablet press instrumented with force and displacement sensors. Furthermore, the compression and compaction behaviour of binary powder mixtures, consisting of an active pharmaceutical ingredient (API) and an excipient, were investigated under systematic variation of the API concentration. Additionally, a new process function for the description of the in-die compression curves was developed based on mechanistic considerations and their applicability for pharmaceutical powders with different deformation behaviour as well as for binary powder mixtures, was examined. It is found that the new process function is applicable for all model materials and well suited for the derivation of characteristic material parameters. The characteristic material parameters can facilitate formulation and process development.

References

[1] F. Podczeck, M. Sharma, The influence of particle size and shape of components of binary powder mixtures on the maximum volume reduction due to packing, Int. J. Pharm. 137, 41–47.

[2] K.C. Tye, C. Sun, G.E. Amidon, Evaluation of the effects of tableting speed on the relationships between compaction pressure, tablet tensile strength, and tablet solid fraction, J. Pharm. Sci. 94, 465–472.

[3] E.N. Hiestand, J.E. Wells, C.B. Peot, J.F. Ochs, Physical Processes of Tableting, J. Pharm. Sci. 66, 1977, 510–519.

[4] H. Leuenberger, B.D. Rohera, Fundamentals of Powder Compression. I. The Compactibility and Compressibility of Pharmaceutical Powders, Pharm. Res. 3, 1986, 12–22.

[5] P. E. Wray, The Physics of Tablet Compaction revisited, Drug Dev. Ind. Pharm. 18, 1992, 627–658.

[6] D. Train, An investigation into the compaction of powders, JPP. 8, 1956, 745–761.

[7] S.N. Vaidya, G.C. Kennedy, Compressibility of 18 Molecular Organic Solids to 45 kbar, J. Chem. Phys. 55, 987–992.

[8] S. Pedersen, H.G. Kristensen, Change in crystal density of acetylsalicylic acid during compaction, S.T.P. Pharm. Sci. 4, 1994, 201–206.

[9] E.V. Boldyreva, High-pressure diffraction studies of molecular organic solids. A personal view, Acta Cryst. A64, 218–231.

[10] S.K. Dwivedi, R.J. Oates, A.G. Mitchell,

[11] J.S.M. Garr, M.H. Rubinstein, An investigation into the capping of paracetamol at increasing speeds of compression, Int. J. Pharm. 72, 117–122.

[12] A.R. Cooper, L.E. Eaton, Compaction behaviour of several ceramic powders, J. Am. Ceram. Soc. 45, 1962, 97–101.

[13] R.W. Heckel, Density-Pressure Relationships in Powder Compaction, Trans. Metall. AIME. 221, 1961, 671–675.

[14] K. Kawakita, K.-H. Lüdde, Some considerations on powder compression equations, Powder Technology. 4, 61–68.

508477 Tablet Disintegration: Advanced Methodology Development

Abstract:

Nowadays, most of the drugs are introduced into the organism in the form of orally administered tablets. It is commonly known that composition, excipients and manufacturing process significantly affect behaviour of the tablets when exposed to aqueous media. significantly affect behaviour of the tablets when exposed to aqueous media. There is a big effort to quantitively and qualitatively describe disintegration, using techniques such as UV-VIS imaging, terahertz time-domain spectroscopy or Raman spectroscopy. In the presented work, there is an overview of used techniques to describe disintegration on binary model tablets made of mixture of ibuprofen and lactose (content of ibuprofen from 0.2 to 0.5 w/w), prepared at different compaction pressures (low, medium and high) are subjected the tests, which should help to identify their dissolution behaviour, mechanical and surface properties, wettability or liquid penetration.

Certified dissolution tests showed that the composition is significant in the most cases (there is a step change in dissolution times between 0.2/0.3 and 0.4/0.5 w/w). Further testing in aqueous environment revealed that there are some properties that are dependent on both composition and compaction pressure.

Measurements by sessile drop method provided information about soaking time and contact angle. Contact angle experiments shows surface tendency to get more hydrophobic with ibuprofen increment. This fact also has minor role in soaking time of liquid to tablet, where major influence is done by manufacturing pressure and thus porosity of tablet.

Experiments based on Texture Analysis exploring the structural properties showed dependence both on compaction pressure and composition as well as on used solution. Acidity plays significant role in tablet strength when there is risk of side reaction with excipient. Overall structure weakening was observed independently on used solution.

Magnetic Resonance Imaging was used for quasi-online monitoring of tablet behaviour in different solution. From experiment can be shown that both higher ibuprofen content and higher compaction pressure slows down disintegration and dissolution processes as well as local saturation of component in solution. There is a surface erosion of tablet observed, when exposed to dissolution medium. Through MRI it can be observed different tablet behaviour such as laminating, disintegration to clusters or dissolution of whole tablet.

The roughness of the tablet surface depends only on the composition and can be evaluated via Scanning Electron Microscopy and subsequent image analysis. Image analysis showed progressive lactose washing out with revealing ibuprofen crystals.

Chemical composition of the tablets can be determined using FTIR microscopy or imaging, where not only the amount but also the morphology of the primary particles on the tablet`s surface can be detected. The size of the primary ibuprofen particles was determined and the differences between two ibuprofen suppliers were clearly detected.

Dissolution and disintegration tests in discriminative conditions showed significant decrease in dissolution rate of ibuprofen, when lactose dissolution was prevented. However, there is an interesting discovery that the tablets with 0.2 w/w ibuprofen are disintegrating also in the case when neither component is dissolving.

These findings lead to the idea of the disintegration can be driven by affinity of the hydrophobic substance to the solvent, which facilitates water penetration and subsequent disintegration and that disintegration occurs due to weakening of inter-particle forces in presence of solution.

509395 Tablet Coating By Injection Molding Technology: Characterization of Coating Materials for Optimal Process and Product Performance

Abstract:

Purpose

ablet coating is one of the most commonly used pharmaceutical unit operations in tablet manufacturing. It renders possible advantages of taste and odor masking, physical and chemical protection, product differentiation, and elegant look to the finished tablet products. Spray coating is the most frequently used film coating process which involves organic and aqueous solvent based polymer systems. The organic solvents can be flammable, costly, and toxic in nature, whereas moisture sensitive drugs cannot be coated with aqueous coating systems. Longer drying time and resulting higher energy costs are other disadvantages of aqueous systems. We have developed a tablet coating process using solvent-free injection molding (IM) technology that can address the limitations experienced by other coating processes. Also, IM technology could be suitable for continuous manufacturing if incorporated with a continuous IM tableting platform. The aims of this research were to 1) optimize coating formulation attributes and process parameters of IM technology, 2) characterize physical properties, process performance and resulting quality attributes of pharmaceutical coating materials, and 3) evaluate the suitability and applicability of IM coating for pharmaceutical tablets.

Methods

Coating formulations (coating polymer and plasticizer in particular ratios) were fed to a vertical, miniature, twin-screw extruder and the generated extrudates were collected through the exit port. Dogbone shaped specimens were created with a microinjector and tested on a universal testing machine. Major mechanical test parameters, like Young's modulus, percentage elongation at the break, toughness, tensile stress at break and tensile strength were determined. A melt flow analysis was conducted using the microinjector with some modifications in the original instrument. Various coating formulations were fed into the heated IM barrel until it was completely filled. A pressure of 600-630 psi was applied by the injection piston onto the coating formulation and the material coming out from heated barrel was then collected in a container to measure the melt flow (g/min). Based on the mechanical properties and melt flow analysis, coating formulations were selected to coat tablets using an in-house built vertical IM machine in a 2-step process. Briefly, the temperature controlled injection barrel was first filled with a coating formulation and the uncoated tablet was placed inside the cavity of the mold inserts. Next, an injection piston applied pressure to the coating formulation and allowed the softened coating material to flow from the barrel to the mold insert. The coating material solidified inside the mold cavities and rendered a smooth coating layer attached to the first half of the tablet surface (step 1). Mold halves were opened, tablet was flipped over and step 2 coating was performed similar as step 1 to obtain fully coated IM tablet. Process parameters (barrel temperature, injection pressure, mold temperature, and cooling time) were optimized for different coating formulations. Coated tablets were stored in various storage conditions and evaluated after 8 weeks to evaluate the coat stability. Drug release from uncoated and coated griseofulvin (GF) tablets was studied, as per GF tablets USP monograph - test 1 dissolution methodology.

Results

Coating formulations suitable for IM process were ranked by evaluating different material properties (Young's modulus, toughness, percentage elongation, and tensile strength/Young's modulus ratio) acquired from the stress-strain analysis. The melt flow property of the coating formulations played a crucial role in IM processing. IM process parameters like injection pressure, barrel temperature, and mold temperature were critical for successful IM coating and were evaluated in depth. Sufficient room humidity (>30% RH) was required while employing polyethylene oxide (PEO) based formulations to avoid immediate cracks, whereas other formulations were insensitive to the room humidity. Tested formulations based on Eudragit E PO and Kollicoat IR did not have the desired tensile properties. Based upon this study, hydroxypropyl peastarch + 30% glycerol, PEO 1,000,000 + 30% PEG and Opadry + 25% glycerol were proven to be viable coating formulations for IM based tablet coating. These formulations had favorable tensile (<700 MPa Young's modulus, >30% elongation, > 95×10^4 J/m^3 toughness) and melt flow (>0.4 g/min) properties, that provided suitable IM coats. These three formulations increased the dissolution time by 10, 15 and 35 minutes respectively (75% drug release) compared to the uncoated tablets (15 minutes). Coated tablets stored in several environmental conditions remained stable to cracking for the studied 8-week time period.

Conclusions

The study confirmed IM as a viable technology for tablet coating. The study serves as a model for product development with specifications of excipients in ranges within the designed acceptance space for optimal product performance.

509487 Quantification of Lubrication and Particle Size Distribution Effects on Tensile Strength and Stiffness of Tablets

Abstract:

We adopt a Quality by Design paradigm to better control the mechanical strength of tablets as a critical quality attribute by understanding the effects of critical process parameters and critical material attributes. To this end, the effect of particle size distribution, lubricant concentration, and mixing time on the tensile strength and stiffness of tablets were studied.

Two grades of lactose, lactose Î±-monohydrate and spray-dried lactose, were selected. Tablets were compressed to different relative densities ranging from 0.8 to 0.94 using an instrumented compactor simulator. We propose a general model, which predicts the elastic modulus and tensile strength envelope that a specific powder can obtain based on its lubrication sensitivity for different particle size distributions. This was possible by introducing a new parameter in the existing tensile strength and elastic modulus models. A wide range of lubrication conditions was explored and the model exhibited a good predictability. The mechanical properties of lactose monohydrate tablets were noticeably dependent on particle size, unlike spray-dried lactose where little to almost no sensitivity to initial particle size was observed. The model is designed in a general fashion that can capture all the possible mechanical integrity behaviors in response to different lubrication conditions and initial particle size. Our model can be extended to all the powders that undergo different deformation mechanisms and is applicable for more complex pharmaceutical formulations.

507254 Evaluating the Influence of Dry Granulation Processing Variables on the Rheological Properties of Granules

Abstract:

Granulation is often employed in the production of pharmaceutical formulations for oral solid dose delivery. It is most often required when constituent materials in a formulation have poor flow properties, enabling these components to be combined to improve process performance and ensure content uniformity.

Dry granulation, based around the use of a roller compactor and integral mill/screen, can have significant advantages over more common wet granulation techniques. Dry granulation can reduce processing costs and can also be used with moisture/solvent/heat sensitive materials. Unlike a wet granulate, the dry granulate does not require additional drying and milling before encapsulation or tabletting.

Dry granulation is not suitable for all types of material, however, there is little research into which properties make a formulation suitable for this process. In addition, there is little to indicate which process parameters produce optimal granulate quality in order to achieve interruption free processing and high quality final products.

In this study an FT4 Powder Rheometer® (Freeman Technology, Gloucestershire, UK) was used to demonstrate how processing parameters influence the properties of a placebo formulation granulate. The granulate was produced using a roller compactor (MiniPactor, Gerteis Maschinen+Processengineering AG, Switzerland) on which the roll gap and compaction force were varied. The dynamic flow, bulk and shear properties of the resulting granulates were then evaluated using the FT4 Powder Rheometer.

The data clearly demonstrates that several rheological properties, known to influence die filling efficiency and Critical Quality Attributes (CQAs) of the resulting tablets, could be controlled by changes in roller compactor settings.

This study emphasises the importance of assessing powder properties using techniques that closely simulate process conditions and how it is possible to identify Critical Process Parameters to optimise a dry granulation process and ensure high final product quality.

507489 Assessment of the Homogeneity of an Alumina Powders Mixture Using Rheological Parameters

Abstract:

This study has been carried out in order to improve the comprehension of powder mixing in the nuclear fuel manufacturing process. Thus, a binary mixture of two Alumina (Al_2O_3) powders, GE15 and CR6, was used to simulate the oxides powders actually involved in the process. Different mixing technics and parameters were tried and compared in terms of homogeneity of the mixtures produced. The concept of intensity of segregation [1] was used for assessing mixture homogeneity through a sampling protocol and classical statistical analysis tools [2]. One of the difficulties of working with these compounds is that, due to the similarities between CR6 and GE15 powders (same chemical composition, same colour, etc.), the direct measure of the composition of a given sample is not possible. Since homogeneity of a mixture is assessed from the variation of the composition between samples, it is important to find an indirect way to measure the composition of an Alumina powders mixture. Therefore, different correlations between measurable parameters (specific surface area, Sauter diameter, cohesion and flow function coefficient) and the composition of the mixtures were found empirically allowing indirect measurement of the composition of a given sample.

The four studied parameters: specific surface area (, measured by nitrogen adsorption using BET method), Sauter diameter (, measured by laser granulometry, dry transportation with air at 0.1 bar), cohesion and flow function coefficient (and , measured by FT4 shear test at 9 kPa on a 10 mL cell), were compared in terms of precision, relevance and feasibility.

It appears that the homogeneity assessment methods involving rheological parameters (cohesion and flow function coefficient) show many advantages compared to the other methods. Not only are they the most accurate correlations found but both parameters are measured simultaneously, meaning that it is possible to apply both methods on the same samples while all the tests (N_2 adsorption, laser granulometry and FT4) are destructive. Those two methods also allow the estimation of the homogeneity at a more interesting scale (around 7 g of powder), indeed, specific surface area and Sauter diameter methods give a homogeneity at a scale of less than one gram which is not relevant for the end-use properties of the final product.

[1] P. Danckwerts, The definition, measurement and some characteristics of mixtures, Applied scientific research. Section A, 3 (1952) 279-296

[2] S. Massol-Chaudeur, H. Berthiaux, S. Muerza, et J. Dodds, A numerical model to identify the structure of a high-dilution powder mixture, *Powder Technol.*, 128 (2002) 131-138

508516 Characterising the Grain Shape: In Search of Size Independent Shape Descriptors

Abstract:

Granular materials contain a large number of grains that differ in size and shape. While the characterisation of the size distribution is typically done using sieve analysis, the characterisation of the shape of the grains remains a difficult task and is one of the major bottlenecks in the characterisation of the granular material. In this study, we attempt to characterise the shape of three different types of grains commonly used in iron and steel making industry, namely pellet, sinter and iron ore lump. While the pellet particles can be considered nearly spherical in shape, imposing the spherical shape on the iron ore lumps and highly irregular sinter particles seems a very poor choice. We consider two different size ranges of pellet particles and four different size ranges for sinter and iron ore lumps. Imaging of the grains is done using an iPhone 7 Plus camera and sufficient number of grains of each type in all the size ranges are considered. We perform the image analysis and obtain various boundary parameters (such as projected area, minor and major axis, perimeter, etc.) of the grains using MATLAB image processing toolbox. In addition, we define two additional boundary parameters to characterise the deviation of the projected area from that of a circle and calculate their values for all the different type of grains considered in this study. Using these length scales, we define a total of nine different shape descriptors and obtain the distribution of these shape descriptors for each type of grain for all the size ranges. Interestingly, for a given type of grains, the frequency distribution curves for these nine shape descriptors turn out to be nearly independent of the size range of the grains. This suggests that the chosen shape descriptors indeed represent properties related to the shape of the grains alone, irrespective of the size of the grains considered.

We investigate the correlation among these shape descriptors by evaluating the values of the correlation coefficients between each pair of the shape descriptors using the Pearson method, Kendall method and Spearman method. Our results indicate that out of the nine shape descriptors considered in this study, five shape descriptors are strongly correlated with each other and only the remaining four shape descriptors need to be considered in characterising the shape of the grains. Out of these four uncorrelated shape descriptors, two shape descriptors, namely the aspect ratio of the grains and the root of form factor, have been considered important by researchers in previous studies. The two other shape descriptors proposed in this study, which are used to characterise the deviation of the projected grain shape from that of a circle, are not measured previously to the best of our knowledge. We compare these four shape descriptors for the pellets, sinter and iron ore lumps having the same sieve size range and show that the shape descriptors for these three different types of grains are different.

We next calculate the values of these four shape descriptors for relatively larger coke particles using 2d image analysis. We also perform 3d image analysis of some of these coke particles and calculate the values of the analogous shape descriptors in 3 dimensions. We investigate if the identified 2d shape descriptors (obtained from 2d imaging) can be correlated to the analogous 3d shape descriptors (obtained from 3d imaging) and explore if it is possible to use the 2d imaging technique to estimate the 3d shape descriptors, thus eliminating the need of 3d imaging.

508543 Investigating Mechanical Properties of Surfactant Films at the Solid-Liquid Interface Using AFM

Abstract:

Corrosion is a well-known problem in the oil and gas sector and also in many other industries. Surfactant based inhibitors have been used for corrosion control in oil pipelines but their inhibition mechanism is still not fully known. Our group is studying surfactant corrosion inhibition mechanism by various characterization techniques (AFM, Electrochemical and Spectroscopic techniques). A better understanding of corrosion inhibition mechanism will lead to the design and development of not only cost-effective but also greener surfactants system for corrosion inhibition purposes.

Atomic force microscopy (AFM) is an effective tool to study the barrier properties of surfactant films at the solid-liquid interface. Determination of mechanical and topographical properties of single and mixed surfactant coatings by atomic force microscopy (AFM) can delineate the role of physical vs. electrochemical barrier by such films in the corrosion inhibition process. . So far, the AFM data, obtained using a closed cell, indicate the effect of surfactant concentration and that of hydrocarbon chains interactions on the mechanical properties of cationic surfactant films. Ultimately, integration of the AFM and electrochemical results will reveal the correlation between the surfactant structures at the solid-liquid interface and their corrosion inhibition efficacy.

508545 Deposition Rate Consequences of the Formation of Multi-Spherule Cluster Aggregates in Gases "Role of Momentum Shielding™"

Abstract:

For engineering flow systems characterized by particulate suspensions in gases, will the formation of *cluster aggregates* (CAs) in the mainstream accelerate deposition rates on immersed targets or confining walls ---or is aggregate formation more likely to suppress capture rates? Recent research on the drag properties of large cluster aggregates (e.g., stimulated in part by the Sorensen (2011) review) has opened the door to answering these important mechanism- (and pressure-) dependent questions. Sub-100 nm "primary" spherules, which may find themselves in even tenuous 'fractal-like' N-spherule clusters, can take advantage of what we call "momentum shielding"---analogous to the drag advantages experienced by birds who elect to fly in formation. We show here that the extent of this "momentum shielding" can be conveniently quantified (in a dimensionless function we call: $S_{mom}(N;Kn_1;\text{aggregate structure})$), and used to predict the deposition-rate consequences of aggregation in aerosol flow systems when the cluster deposition mechanism is dominated by either: i) isothermal convective diffusion (C-D), ii) thermo-phoresis (TP-) or iii) inertial impaction (II). As expected, we show that for any type of aggregate structure (including 'fractal-like'), momentum shielding increases with cluster size (e.g. N) and gas pressure (via the reciprocal Knudsen number based on spherule radius R_1). We also show how S_{mom} can be calculated for the limiting case of a N>>1 random close-packed-D_f=3 cluster ---which is nearly gas impermeable. For fractal-like aggregates, including DLCAs (D_f=1.8) and RLCAs (D_f=2.1) we have recently shown that the abovementioned function: $S_{mom}(N, Kn_1; D_f)$ is actually sufficient to make useful engineering estimates of CA Brownian diffusion coefficients (Rosner and Tandon (2017b), thermophoretic diffusivities (Rosner and Tandon (2017a) and aggregate "stopping times" (Rosner and Tandon (2017c)----each in terms of their previously known values (D_1, $\alpha_T D_1$, and $t_{p,1}$) for isolated individual spherules in the same environment.

This information now enables instructive engineering calculations of the deposition-rate consequences of spherule aggregation to form large cluster aggregates-----i.e. at the same mainstream spherule volume fraction we can now determine how much aggregation will increase or decrease the total spherule acquisition rate on an immersed target. We present here representative results for each of the abovementioned particle transport mechanisms----i.e.: i) isothermal convective-diffusion (C-D), ii) thermo-phoresis (TP-) or iii) inertial impaction (II)----- including a brief account of the consequences of dealing with Kn1-dependent coagulation-'aged' aggregate size distributions (ASDs) in the mainsteam. We summarize our current re-examination of the canonical problem of inertial impaction on a circular cylinder in Re>>1 crossflow (Rosner and Tandon (2017c), dealing now with CA-"projectiles" but in the absence of spherule loss from the much larger target. We conclude by briefly outlining our planned work on the efficient "inversion" of sampled cluster-ASD-data to reconstruct, using the abovementioned $S_{mom}(N,Kn_1;D_f$ information, what ASD must have existed in the mainstream---at least in the absence of appreciable supercritical aggregate break-up.

References:

1. Rosner, D. E. and P. Tandon (2017a). Knudsen Transition Effects on the Thermophoretic Properties of Fractal-like Aggregates; Implications for Thermophoretic Sampling of High-Pressure Flames, Aerosol Science and Technology (in press).

2. Rosner, D. E. and P. Tandon (2017b). Aggregation- and Knudsen Transition Rarefaction- Effects on Particle Mass Deposition Rates by Convective-Diffusion, Thermophoresis or Inertial Impaction; Consequences of Multi-spherule 'Momentum Shielding', Aerosol Science and Technology (in press).

3. Rosner, D. E. and P. Tandon (2017b). Mainstream Aggregation and Polydispersity Effects for Inertially-Dominated Deposition Rates on Cylinders (or Leading 'Edges') in Crossflow"---- Consequences of 'Momentum Shielding' for the Impaction of Large, "Compact" ($2<D_f<3$) Multi-spherule Aggregates, Aerosol Science and Technology (prepared for submission, August 2017).

4. Sorensen, C. M. (2011). The Mobility of Fractal Aggregates: A Review, Aerosol Science and Technology. 45: 765-779.

508609 Experimental and Numerical Studies on the Thermal Diffusivity of PackedÂ Powder Beds

Abstract:

Several industrial processes, such as calcination and drying, involve thermal treatment of particles. In order to achieve efficient processes and good quality products, the thermal properties of the particles and that of the bulk powder must be well defined. Previously, some models, such as the Krischer and Maxwell-Eucken models, have been proposed to relate particle thermal properties to that of the bulk powder. Though these models have been shown to work well, it is not very clear how to account for changes in the packing density, moisture content, and arrangement of particles. In addition, the particle thermal properties of most materials are not readily available, which limits the application of these models. To get a better understanding of the relationship between the particle thermal properties and that of the bulk powder and also to quantify the effects of packing density and moisture content on the bulk powder thermal properties, we run several experiments and discrete element method (DEM) based simulations. In both the experiments and simulations, a pipe is filled with particles where the temperature of the pipe is kept constant. We measured the temperature change in the powder bed and compute the actual thermal diffusivity of the powder bed. We tested several types of powders and glass beads in the experiments. The packing fraction, the moisture content, and size of particles were varied in the experiments. In the simulations, in addition to the packing fraction and particle size, the thermal properties of the particles were also varied. In this talk, we will present the details of the experimental and simulation results.

509556 Determination of Size, Size Distribution and Refractive Index of Artificial and Biological Microparticles

Abstract:

For the approval of clinical studies, it is mandatory to validate the production process of artificial oxygen carriers by reliably specifying the properties of the microparticles used like the newly developed blood substitute HbMP-700 (1-3). In particular, besides shape, size and size distribution, the amount of hemoglobin (Hb) and methemoglobin (metHb) content is essential.

We present a method based on the simultaneous measurement of collimated and diffuse transmission spectra in the visible and ultraviolet range and numerical analysis to infer the wavelength-dependent refractive index, size and size distribution of microparticles.

For the validation of the method, we used aqueous suspensions of monodisperse polystyrene microspheres of known concentration, determined by flow cytometric measurements. The related data analysis is carried out for synthetic and experimental data. For homogeneous spheres, Mie theory is applied to compute the solution for the scattering of a plane electromagnetic wave by a particle and to calculate the extinction cross section C_{ext}, i.e. the shadow cast by the particle as a function of the vacuum wavelength λ, the particle size, and the complex refractive index $n(\lambda)$. For dilute suspensions of particles, the measured quantity is an ensemble-averaged extinction cross section C_{ext} . This average is obtained by integrating over the particles' size distribution, which may not be known a priori. From the resulting spectral extinction cross sections, the complex refractive index and the mean and standard deviation of the size distribution can be inferred by means of nonlinear least-squares optimization. In order to reduce the number of free parameters of the problem to a feasible amount, $n(\lambda)$ is represented in a set of harmonic-oscillator-like resonance functions. The refractive index of polystyrene microspheres obtained is in good agreement with literature values for the bulk material, the mean and standard deviation of the distribution of particle *size D* are determined with an uncertainty in the range of a few nanometers. For example, for microspheres of mean diameter mean(D) = 2μm and distribution width std(D) = 9.4nm the uncertainty of the mean is $u\{\text{mean}(D)\}$ = 1.2nm and the uncertainty of the distribution width amounts to $u\{\text{std}(D)\}$ = 0.4nm.

In a second step, we applied the method to characterize HbMP-700 microparticles, approximately featuring a prolate spheroidal shape with 700nm length and 400nm width. Besides shape, size and size distribution, the amount of hemoglobin (Hb) and methemoglobin (metHb) content is essential. The relative concentrations of oxyHb, deoxyHb, metHb are accessible from spectral extinction and diffuse transmission measurements, since the different Hb variants exhibit different wavelength-dependent refractive indices. The method to infer size and refractive index described above can be applied to these non-spherical particles, since the deviation of their light scattering properties from those of spheres turns out to be small. The results based on the optical method will be compared to alternative procedures for HbMP particle characterization with respect to their size and size distribution, i.e. electron microscopy, centrifugation and dynamic light scattering. In addition, x-ray spectrometry will be used to detect the ratio between Fe^{2+} (oxy/dexyHb) and Fe^{3+} (metHb) as well as the iron concentration.

References

1) Novel hemoglobin particles--promising new-generation hemoglobin-based oxygen carriers. Bäumler H, Xiong Y, Liu ZZ, Patzak A, Georgieva R. Artif Organs. 2014 Aug;38(8):708-14. doi: 10.1111/aor.12331.

2) Nonvasoconstrictive hemoglobin particles as oxygen carriers. Xiong Y, Liu ZZ, Georgieva R, Smuda K, Steffen A, Sendeski M, Voigt A, Patzak A, Bäumler H. ACS Nano. 2013 Sep 24;7(9):7454-61. doi: 10.1021/nn402073n.

3) Hemoglobin-based oxygen carrier microparticles: synthesis, properties, and in vitro and in vivo investigations. Xiong Y, Steffen A, Andreas K, Müller S, Sternberg N, Georgieva R, Bäumler H. Biomacromolecules. 2012 Oct 8;13(10):3292-300. doi: 10.1021/bm301085x.

517131 FlowCam NanoÃ,Â® Provides Counts, Sizes and Images of Nano-and Microparticles: Application to a Therapeutic Protein Pumping Study

Abstract:

Sub-visible particle characterization is becoming a more important method in assessing drug substance and drug product stability. However, the current USP protocol does not require any characterization of particles below 10 Î¼m. This greatly hampers our ability to properly assess protein stability as a function of particulate formation and obscures most sub-visible particles and various morphological properties.

In this study, we assess the capabilities of the new FlowCam Nano from Fluid Imaging Technologies, in conjunction with a pumping study that focuses on particulate formation generated during fill operations. Initial experiments show that the new FlowCam Nano can generate highly resolved images of nanoparticulates, allowing the simultaneous assessment of particulate morphology for nano- and microparticulates. This is extremely useful, as currently unpublished data (for the pumping study) show that as a function of post-pumping agitation, a significant increase in microparticle concentration coincides with a significant depletion of the nanoparticle concentration. This suggests that agglomeration (and probably not nucleation) is the more likely mechanism for larger microparticulate formation.

The ability to characterize the nano- and microparticulate morphology simultaneously, will allow us to better assess the cause of particulate formation (such as observing the transition of nanoparticles to microparticles and seeing if specific morphological properties of larger nanoparticulates are conserved in microparticulates). In addition, the FlowCam Nano is also sensitive enough to assess differences in the colloidal stability of a drug product in different formulations. During filling operations, and accelerated degradation studies, FlowCam Nano showed the differences in size distributions of IVIG in PBS and 0.25 M glycine buffer. Since IVIG is less colloidally stable in PBS, the size distribution varies significantly more per size bin and the overall mean effective spherical diameter is also ~ 600 percent greater after 24 hours of agitation in comparison to the size distribution data of IVIG in glycine. Thus, Flow Cam Nano has capabilities that are very useful in assessing drugs in the substance (e.g. optimizing drug formulations) and product (e.g. assessing particulates post fill & finish operations) phases.

Numerical and experimental estimation on the normal and tangential capillary bridge force adhered to two spheres

Kazuo Murase*, Keisuke Arai, Takato Ootsuka, Daiki Sakamoto and Futa Egawa,
Department of Applied Chemistry, Chuo University
1-13-27, Kasuga, Bunkyo-ku, Tokyo,1128551, Japan
**hmurase@kc.chuo-u.ac.jp*

Introduction

Liquid bridges extensively occur as a binder liquid adhered to solid surfaces in a variety of engineering applications such as crystal growth, ink-jet printing, material processing and powder technology. Especially the strength of adhesion and cohesion of liquid bridges adhered to particles contribute to the consolidation of an agglomerate in wet granular materials. Estimation of the attractive or repulsive force between particles is important in controlling kinetics of agglomeration and handling of granular materials. Characteristics of a liquid bridge e.g. the equilibrium bridge geometry [1], stability [2], static and dynamic binding force [3-6], rupture energy [7,8], and wetting hysteresis [9] etc., have been experimentally and numerically investigated.

The liquid distributions within a particles bed are classified into four types of categories, i.e. pendular, funicular, capillary and slurry states, according to the liquid contents [10]. The static normal capillary force at pendular liquid state have elaborately been investigated as Fisher [11], toroidal model [8] and the exact solution of Yung-Laplace equation [12]. The dynamic normal capillary forces [13] are experimentally estimated with lubrication theory. The tangential force of a liquid bridge [14] also contributes to the primary factor for the granular consolidation with the shear stress of wet granular materials but have not yet substantially been investigated.

This work aimed to estimate experimentally and numerically the normal and tangential capillary force of a static and dynamic liquid bridge adhered to two spheres. The asymmetry static liquid bridge with the shear stress demand for the exact solution of Young-Laplace equation for the equilibrium liquid drop including the effect of gravity.

1. Theory

1.1 Formulation of a asymmetry static liquid bridge

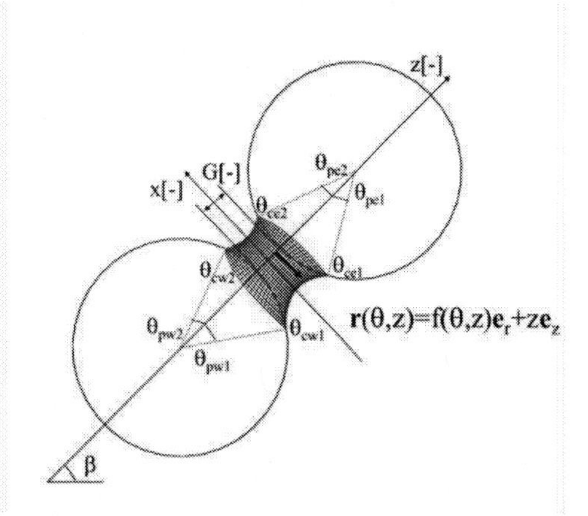

Figure 1 Geometry of a liquid bridge adhered to two spheres with arbitrary inclination in cylindrical coordinate system.

The asymmetry static bridge shape demands for the exact solution of the Young-Laplace equation due to the limit of the toroidal approximation. A prevalent geometry of a liquid bridge adhered to two spheres is drawn in Figure 1. The dimensionless position vector of a liquid surface in cylindrical coordinates is described as $\mathbf{r}(\theta,z) = f(\theta,z)\mathbf{e}_r + z\mathbf{e}_z$ divided by spherical radius, R', which $\mathbf{e}_r$, $\mathbf{e}_z$ are unit basis vectors in r-,z-directions. The symbol (') means the dimensional value. Fig.1 indicates the two dimensional front view with four half-filling (θ_{Pw1}, θ_{Pw2}, θ_{Pe1}, θ_{Pe2}) and contact angles (θ_{Cw1}, θ_{Cw2}, θ_{Ce1}, θ_{Ce2}).

The unit outward normal vector of a bridge surface, $\mathbf{n}_L$, is presented as follows [15],

$$\mathbf{n}_L \equiv \frac{r_\theta \times r_z}{|r_\theta \times r_z|} = \frac{f\mathbf{e}_r - f_\theta \mathbf{e}_\theta - ff_z\mathbf{e}_z}{[f^2(f^2+f_z^2)+f_\theta^2]^{1/2}} \quad \ldots(1)$$

,where low suffixes (r,θ,z) describe partial derivatives.

The dimensionless effective potential energy, $E=E'/\sigma R'^2$ with Lagrangian multiplier, $K=K'/\sigma R'^2$, on the constrained condition of a constant volume, $V=3V'/4\pi R'^3$, is described as follows [16],

$$E = \int_{-\pi/2}^{\pi/2}\int_{z_{min}}^{z_{max}} 2\left[f^2\left(f^2+f_z^2\right)+f_\theta^2\right]^{1/2} d\theta dz + \frac{Bo}{2}\int_{-\pi/2}^{\pi/2}\int_{z_{min}}^{z_{max}} (z\sin\beta + x\cos\beta) f^2 d\theta dz + K\left[V - \frac{1}{2}\int_{-\pi/2}^{\pi/2}\int_{z_{cp1}}^{z_{cp4}} f^2 d\theta dz\right] \quad \ldots(2)$$

Bond number (Bo) is a dimensionless number defined by $(\rho_L-\rho_G)gR'^2/\sigma$, where is the ratio of gravitational force to surface tension force equally as Etovoes number (Eoe).

The integral domain of a bridge surface is determined by the trapezoidal section of the upper and lower surface lines, [$z_{w2}<z<z_{e2}$ at $\theta=\pi/2$, $z_{w1}<z<z_{e1}$ at $-\pi/2$] and contact lines, [$z_{w1}<z<z_{w2}$, $z_{e1}<z<z_{e2}$ at $-\pi/2<\theta<\pi/2$] with four contact points settled by four half-filling angles ($\theta_{Pe1},\theta_{Pe2},\theta_{Pw1},\theta_{Pw2}$) in Fig. 1. Two contact lines on the upper and lower spheres were approximated by ellipse lines with parametric aspect ratios, which pass through respectively pairs of two contact points settled with half filling angles, ($\theta_{Pw1},\theta_{Pw2}$) and ($\theta_{Pe1},\theta_{Pe2}$). Filling angles, ($\theta_{PW}(\theta)$, $\theta_{PE}(\theta)$) on the lower and upper spheres have variable properties along the circumferential direction. The formulation of a static bridge shape at equilibrium state results in the extremum problem, which corresponds to the condition that the first order variation equals to zero, $\delta E=0$.

1.2 Estimation of the capillary force of a liquid bridge

The dimensionless mean curvature (H) and contact angles (θ_c) of a bridge surface are respectively calculated by eqs. (3),(4) once the outline of a liquid bridge is numerically evaluated.

$$-2H = \nabla_s \cdot \mathbf{n}_L = \left[(\alpha^2 + f_\theta^{\,2}) - f(f^2 + f_\theta^{\,2})f_{zz} - f(1+f_z^{\,2})f_{\theta\theta} + 2ff_z f_\theta f_{\theta z}\right]/\alpha^3 \quad \ldots(3)$$

$$,where\ \ is\ \ \alpha^2 = f^2(1+f_z^2) + f_\theta^2$$

$$\theta_c(\theta) = \cos^{-1}(\mathbf{n}_s \cdot \mathbf{n}_L) \quad \ldots(4)$$

,where contact angles also depend on the circumferential angle.

According to the Young-Laplace equation, the dimensionless pressure difference, $\Delta P = \Delta P'R'/\sigma$, across a liquid surface is related with the mean curvature of the surface.

$$\Delta P = (-2H) \quad \ldots(5)$$

The capillary force on a liquid bridge adhered to two spheres must be evaluated only on the contact line as known by the boundary method due to unspecified neck radius of a bridge and asymmetric condition. Dimensionless averaged capillary forces, $F_{Ci} = F'_{Ci}/(2\pi R'\sigma)$ on the upper and lower sphere (i=1,2 respectively correspond to lower, upper sphere) are presented as follows by three terms, which correspond to the surface tension, hydrostatic pressure and buoyancy terms along the longitudinal angle (θ). The sign of the capillary force set to be positive for attractive direction.

$$F_{ci} = \frac{1}{\pi}\int_{-\pi/2}^{\pi/2} \sin(\theta_{Pi})\sin(\theta_{Pi}+\theta_{Ci})d\theta - \frac{1}{2\pi}\int_{-\pi/2}^{\pi/2} \sin^2(\theta_{Pi})\Delta P d\theta \mp \frac{2}{3}\mathrm{Bo}\cdot V_{bi} \quad \ldots(6)$$

, which V_b is sphere volume submerged in a bridge

Finite Element Method is applied for the solution of the $\delta E=0$ equation. Isoparametric elements of a biquadratic polynomial shape function were selected for the integral domain and unknown radius. The set of non-linear algebraic equations for unknown radius at element nodes and Lagrangian multiplier was resolved by the Newton-Raphson method [9].

2. Experiment

Tensile strength of a static and a dynamic pendular liquid bridge

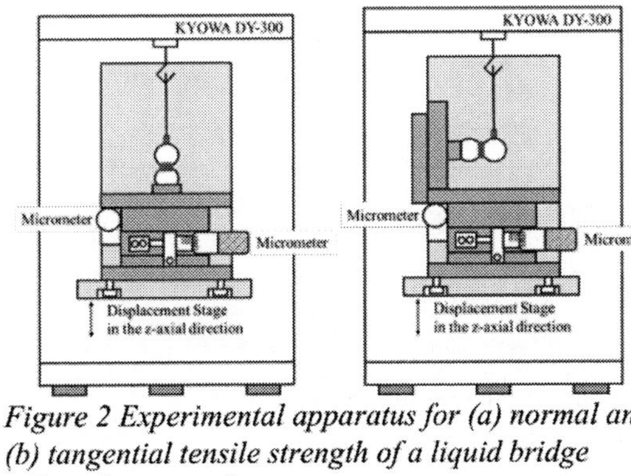

Figure 2 Experimental apparatus for (a) normal and (b) tangential tensile strength of a liquid bridge adhered to two spheres

Figure 2 shows the surface tensiometer (Kyowa interface Science Co., Ltd. DY-300) for measuring (a) the normal and (b) the tangential tensile strength of a water liquid bridge adhered to two stainless steel spheres (radius 3.0mm). The rod attached with one sphere is hanged up to the tensiometer and the other sphere fixed on the displacement stage is moved upward and downward by the tensiometer. The discrete and continuous displacement of the stage are performed for the normal and tangential deformation of a liquid bridge for the static and dynamic states. Two micrometers adjust the appropriate position of the lower spheres for the upper sphere. Water droplets are poured into the spheres aperture and substantially wetted by the displacement of the stage. The tensile strength at static and dynamic state and snapshots of liquid droplets by the digital camera and movie are recorded. The outlines of droplets are evaluated by a digitizer and estimate the half filling angles ($\theta_{Pe1}, \theta_{Pe2}, \theta_{Pw1}, \theta_{Pw2}$) and contact angles ($\theta_{Cw1}, \theta_{Cw2}, \theta_{Ce1}, \theta_{Ce2}$) at four contact points. The sphere configuration (radius, gap), the droplet volume, the half filling angles and the contact lines on spheres initiate parameters of numerical simulation.

3. Results and discussions

3.1 The normal tensile strength of a liquid bridge adhered to two vertical spheres.

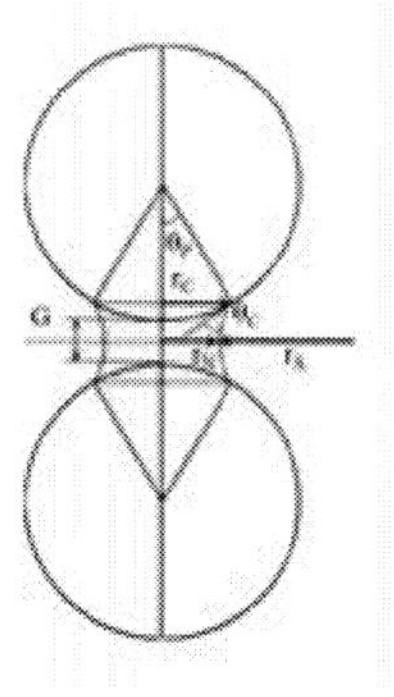

Figure 3 Geometry of toroidal model

The symmetric geometry of a liquid bridge adhered to two vertical spheres without gravity has been extensively investigated by toroidal approximation in Figure 3. The boundary method overestimates the capillary for of a liquid bridge rather than the gorge method but is feasibly and distinctly defined for the asymmetric bridge outline. Symmetric condition simplifies eq.(3) for the mean curvature and eq.(6) for the capillary force into eq.(7),(8) [7,16,17].

$$-2H = \nabla_s \cdot \mathbf{n}_L = \frac{1}{\alpha}\left(\frac{1}{f} - \frac{f_{zz}}{\alpha^2}\right), \ where \quad \alpha^2 = 1 + f_z^2 \qquad \ldots(7)$$

$$F_{Boundary} = r_C \sin(\theta_{Pi} + \theta_G) - \frac{r_C^{\ 2}}{2}(\Delta P) \mp \frac{2Bo}{3} V_{bi}, \ \ r_C = \sin\theta_P \qquad \ldots(8a)$$

$$F_{Borge} = r_N - \frac{r_N^{\ 2}}{2} \sin\theta_{cpi}(\Delta P) \mp \frac{2Bo}{3} V_{bi} \qquad \ldots(8b)$$

,where $r_N - r_A = \dfrac{A}{\tan(\pi/2 - \theta_P - \theta_C)} + \sin\theta_P, r_A = \left[A^2 + \{\sin\theta_P - (r_N - r_A)^2\}\right]^{1/2} A = G/2 + 1 - \cos\theta_P$

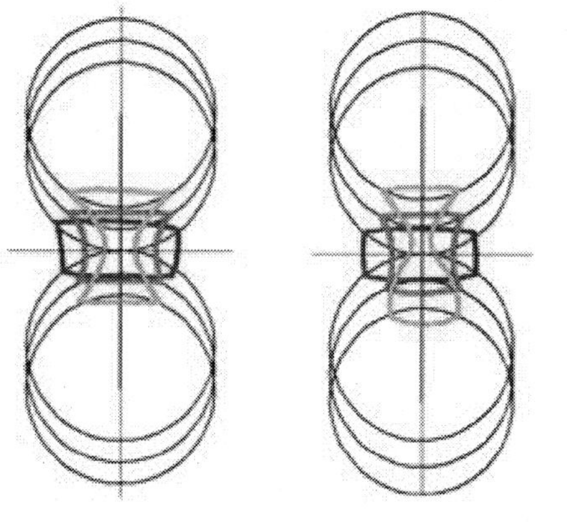

(a) Static state G=0, 0.47, 0.91 (b) Dynamic state G=0, 0.59, 1.16

Figure 4 Experimental droplet outlines at static and dynamic state, Bo=1.21, V=0.088

Figures 4 show symmetric outlines of liquid bridges around the vertical axis respectively at the static and dynamic state. Static state results in the shorter rupture gap and the larger dimensionless normal tensile strength from Figures 5a. Sphere gaps at the peak static capillary force depend on liquid volume. The contact angle on the upper sphere corresponds to the advancing contact angle at the higher Bo number due to gravity effect.

Fig. 4b indicates the downward transition of a liquid immediately before the rupture due to the wettability a liquid. The toroidal approximations over-estimate experimental data at the lower sphere gap but the sphere gaps at the peak dynamic capillary force depend on the velocity of stage displacement. The velocity reduces the peak value of dynamic capillary due to the downward transition of a liquid in Fig. 5b. The stepwise profiles of capillary force are due to the movement sensitivity of tensiometer for the stage displacement.

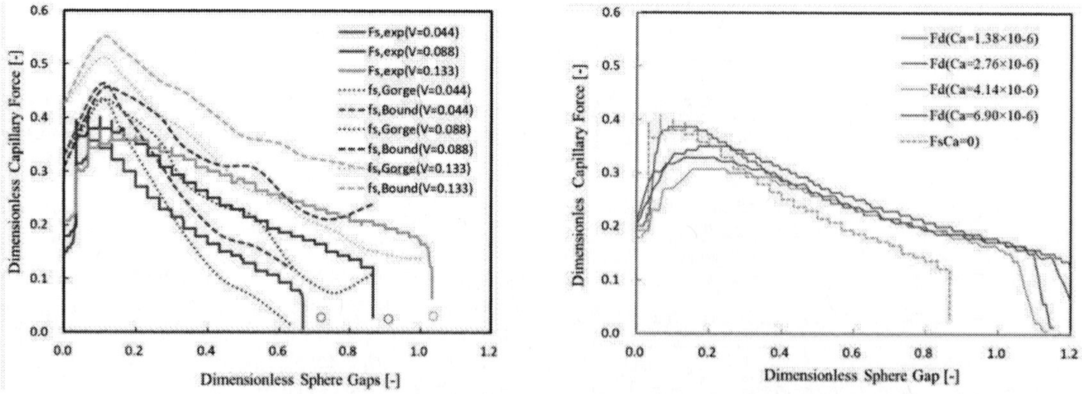

(a) at the static state (b) at the dynamic state

Figure 5 Effect of sphere gap on dimensionless capillary force

3.2 Estimation of boundary condition at contact line by LBM

The Young's equation describes the force balance among three interface tensions (solid-liquid, solid-gas and liquid gas) at three phase contact point and also defines the relationship between interface tension and contact angle. The solution of the Young-Laplace equation at the static state demands for one unknown boundary condition of either a contact angle or a contact line. The simulation for the wettability on three dimensional contact line demands for the elaborate process. Our experiment determines the contact lines on spheres under the constant liquid volume and is utilized for the boundary condition. Besides the perfect wettability the smooth transition of three phase contact lines is unexpected due to the advancing and receding contact angle. Lattice Boltzmann Method (LBM) is expected to estimate the boundary condition at static and dynamic state. The geometry of a liquid bridge is numerically estimated by FEM with half-filling angles predicted by 2 dimensional Shan & Chen model of LBM, which is used by D2Q9 model for transition of distribution function. The estimation for the liquid geometry by FEM has been verified. Figures 6 show the predicted liquid outlines by LBM and FEM. at a given liquid volume.

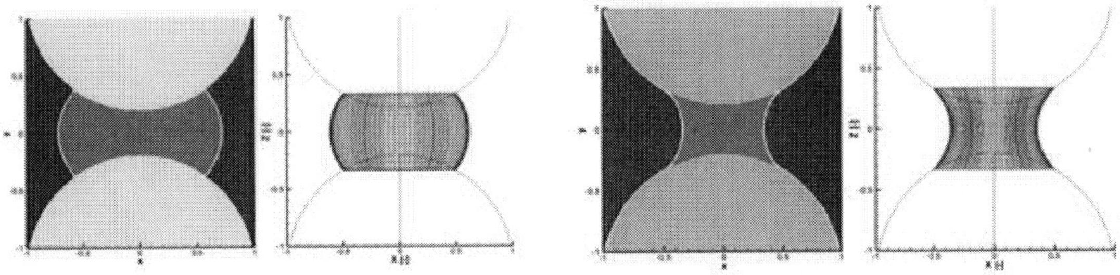

(a) Repulsive interparticle force (b) Attractive interparticle force

Figure 6 Snapshots of liquid geometries at static state by LBM and FEM

Figures 7 show snapshots of dynamic liquid outlines at a given liquid volume up to the rapture state by LBM. The transition of boundary conditions, i.e. filling and contact angles, is smoothly described. Gravity and imperfect wettability, i.e. advancing and receding contact angles, are excluded in Figs. 7 but the effect of gravity on the liquid outline is avairable.

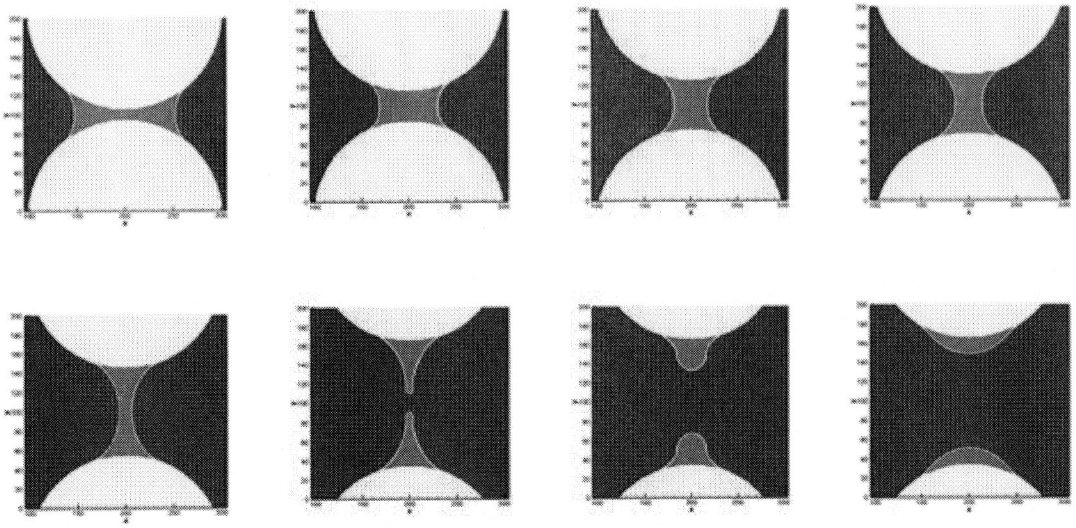

Figure 7 Snapshots of snapshots of dynamic liquid outlines at a given liquid volume

3.3 The tensile strength of a liquid bridge with shear deformation.

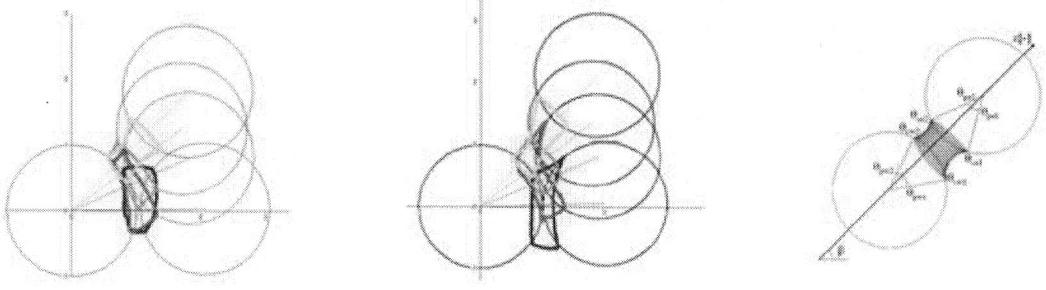

(a) Static state (G=0, 0.59, 0.98) (b) Dynamic state (G=0, 0.59, 0.98)

Figure 8 Liquid outlines at static and dynamic state (Bo=1.21, V=0.088)

Definitions of half-filling and contact angles.

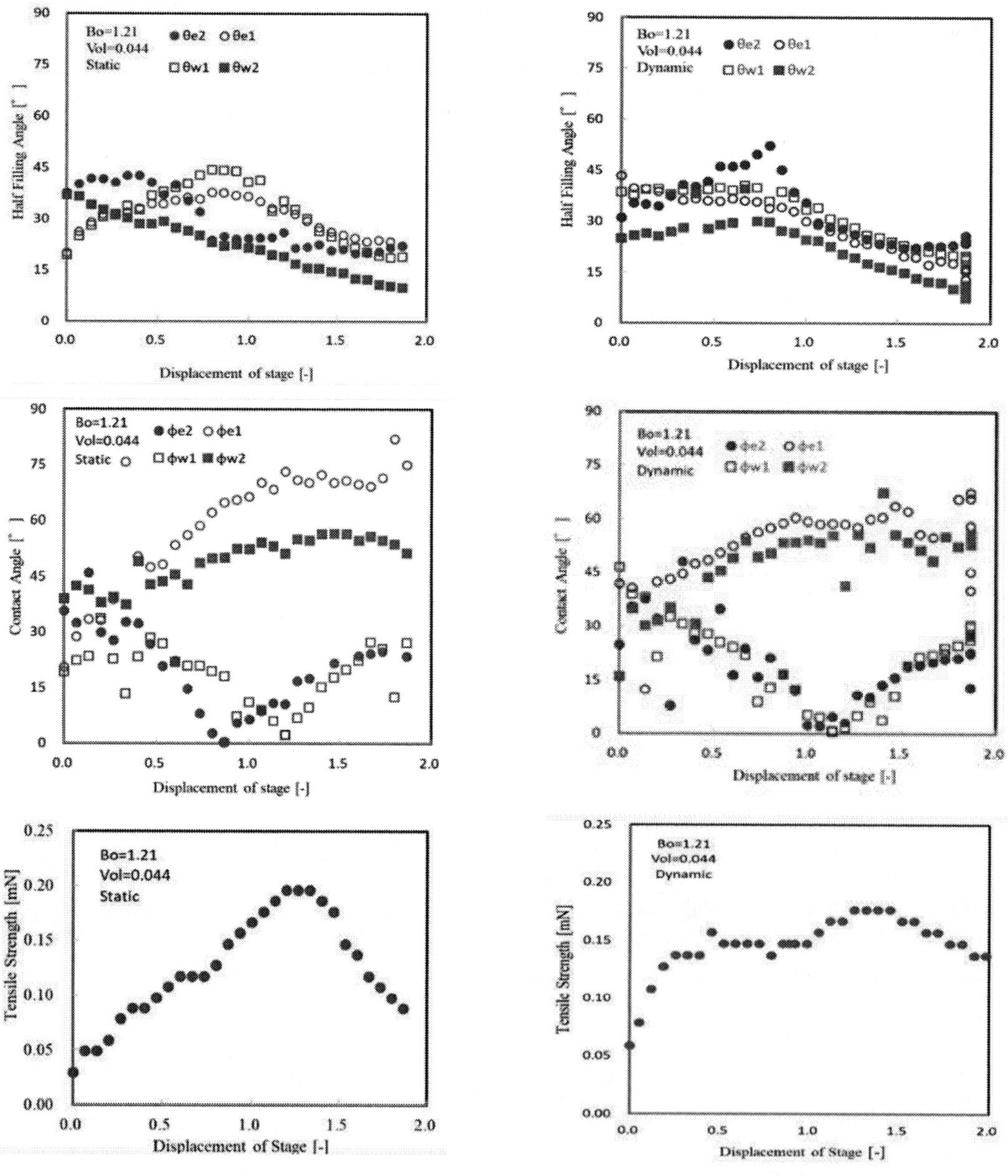

(a) Static state (b) Dynamic state

Figure 9 Profiles of half-filling angles, contact angles and tensile strength , Bo=1.21, V=0.088

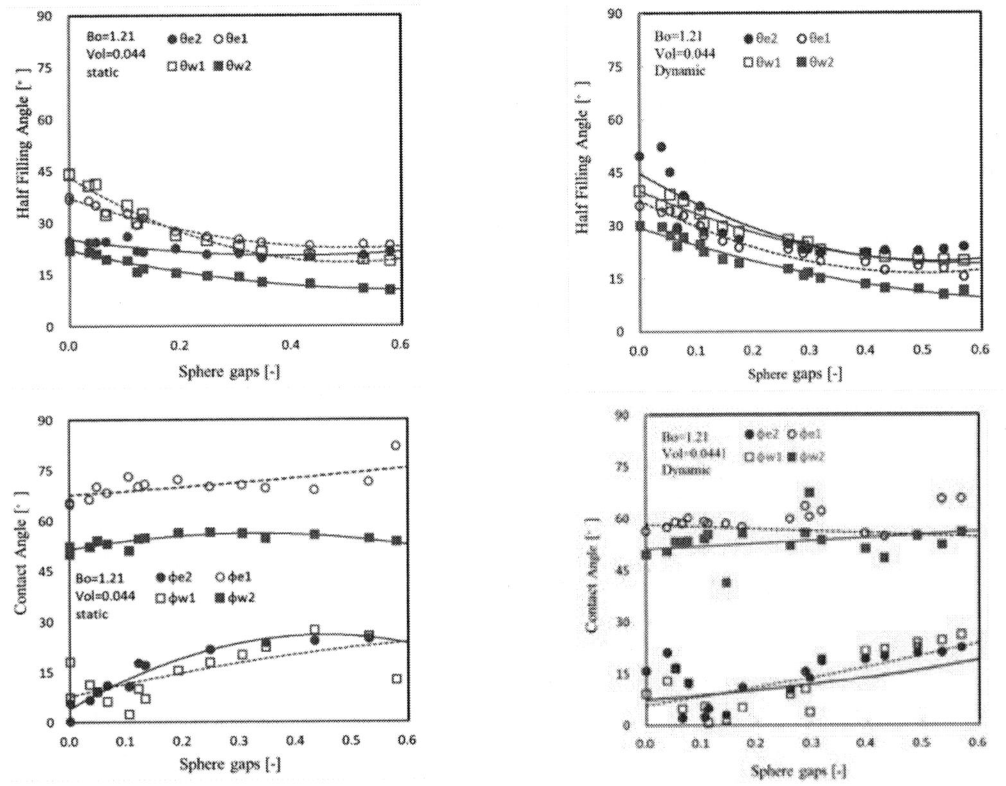

(a) Static state (b) Dynamic state

Figure 10 Profiles of half-filling angles, contact angles with sphere gaps , Bo=1.21,V=0.088

Figures 8 show four types of liquid outlines at static and dynamic state. Magenta and green plots are drawn respectively at the extreme values of half filling angles, contact angles and tensile strengths in Figures 9. During the lower stage displacement two spheres keep at the contact state and the half-filling angles reached to the peak values. Half-filling angles sifted with displacements of a stage have the peak values except the static half-filling angle at the top point of the lower sphere (θ_{Pw2}), which monotonously decrease with stage displacement. The increase and decrease of contact angle respectively corresponds to the advancing and receding contact angle. The contact angles at the bottom point of the lower sphere (θ_{Cw1}) and at the top point of the upper sphere (θ_{Ce2}) transit from the receding to the advancing contact angle due to the infinitesimal values. Figures 10 redraw the half-filling and contact angles with respect to sphere gap. Contact angles contribute to the peak tensile strength, which is measured as resultant force (F_T) of normal (F_n) and tangential (F_t) tensile strength in Figure 11. The normal tensile strength conventionally corresponds to the capillary force. The stage displacement at the peak tensile strength minimize the sum of half-filling and contact angle due to the positive hydrostatic pressure across liquid surface. The larger dynamic tangential strength occurs during the sphere contacting due to the asymmetric deformation. The adhesion force of liquid to sphere at the static state contribute to the larger peak tensile strength.

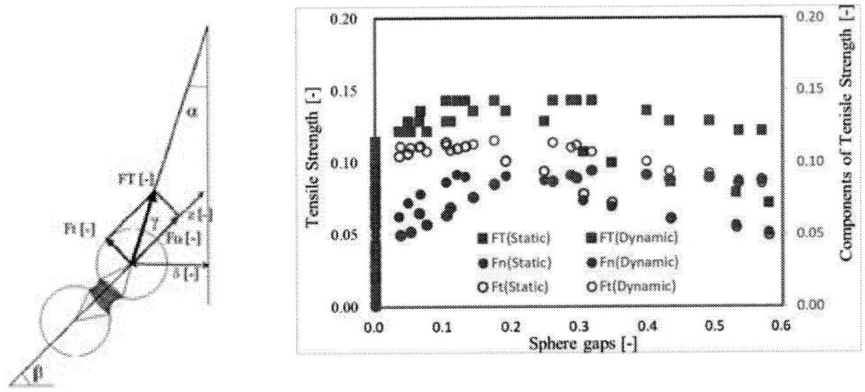

Fig.11 Normal and tangential components of tensile strength at static state

3.4 Numerical simulation at the static state

The solution of Young-Laplace equation demands for one unknown necessary condition on the contact point, i.e. half-filling or contact angle. Liquid surfaces simulated by FEM are drawn with solid lines in Figures 12. The simulation reveals the asymmetric liquid geometry with half-filling and contact angles along longitudinal angle. Disagreements of liquid interfaces results from the approximation of contact lines on spheres. The tangential strength influences depend on the mean curvature. Especially the second figure at the peak half-filling angle has the negative hydrostaic pressure in the direction of relative sphere displacement and positive pressure in the oppsite direction.

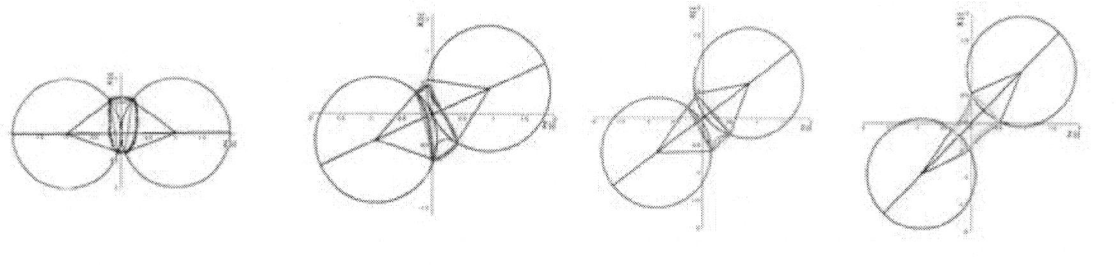

At the initial geometry   At the peak halffilling angle   At the peak tensile strength   At the maximum shpere gap

Figure 12 Numerical estimation at the static state

Figure 13 displays transitions of measured normal tensile strengths ($F_{exp,n}$) and simulated capillary force (F_{cal}) to sphere gap. The initial droplet is attached to the above aperture in the direction of sphere movement due to gravity. The stage displacement drifts parts of liquid volume into the lower aperture of spheres. The sphere on displacement stage keep contacting with the other sphere hang up to the tensiometer up to the peak tensile strength. Fcal and FcalB are numerically estimated by outlines of liquid surfaces respectively with Young-Laplace equation and boundary toroidal model. The sign of hydrostatic pressure with deformation of liquid droplets results in the repulsive capillary force. The normal tensile strength corresponds to the capillary force during the range of symmetric liquid bridge. The angle of beta is defined by the inclination of center distance of a sphere pair. The lower normal tensile strength than the capillary force results from the orthogonal component of capillary force at the low range of beta.

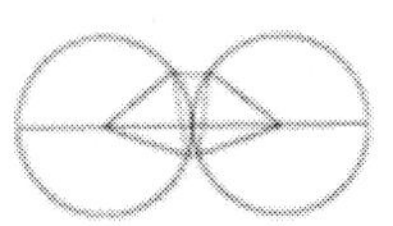

The initial geometry

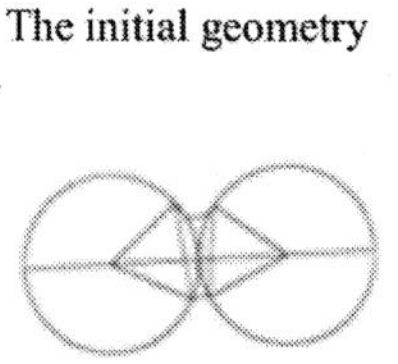

at the peak halffilling angle

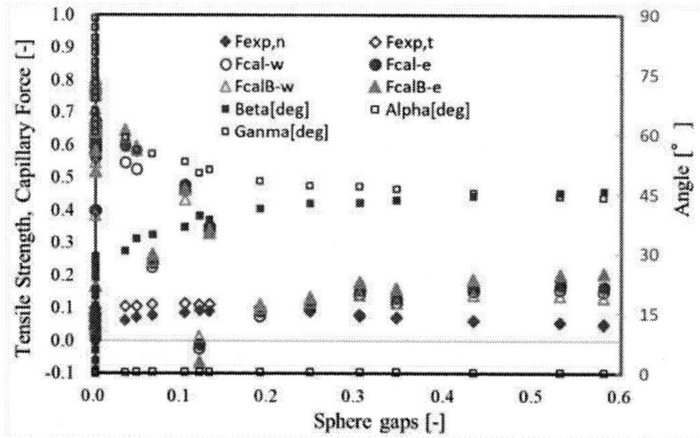

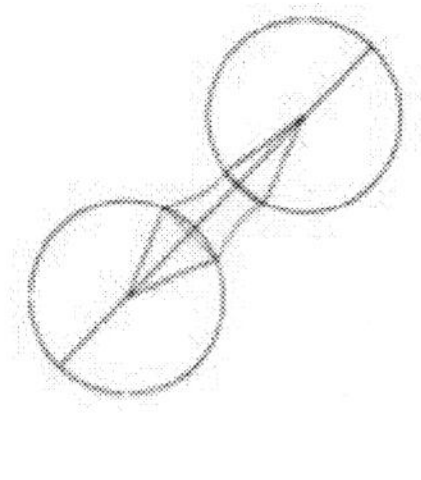

at the maximum shpere gap

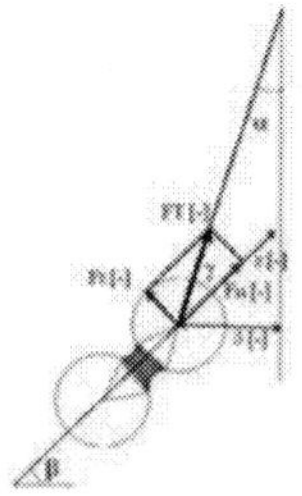

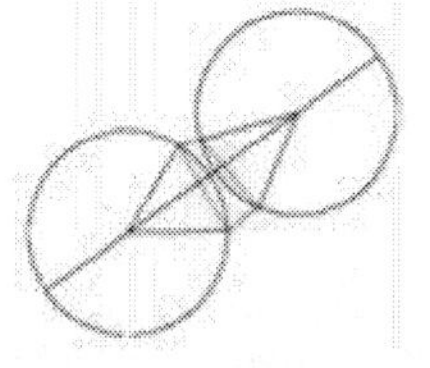

At the peak tensile strength

At the repulsive state

Figure 13 Normal tensile strength and numerical capillary force

Conclusions

Liquid bridges adhered to particles are deformed with the normal and tangential stress in wet granulator. The interparticle force with asymmetric droplet by shear stress must be estimated by the three dimensional solution of Young-Laplace equation. A surface tensiometer and numerical simulation contribute to the normal and tangential tensile strength of liquid bridge adhered to two spheres.

1. The normal strength of symmetric liquid bridge with a surface tensiometer is verified by the conventional toroidal model and the solution of Young-Laplace equation.
2. Sphere gaps at the peak capillary force in the vertical sphere configuration depend on liquid volume and displacement velocity of sphere.
3. The lattice Boltzmann Method is expected to numerically predict the boundary conditions on the contact lines, which are half-filling angle and contact angle at the static and dynamic state.
4. The tensile strength depends on transition of contact angles, which correspond to advancing or receding angles. The larger sum of contact and half-filling angles than the orthogonal angle corresponds to the positive hydrostatic pressure across liquid surface.
5. The normal tensile strength is related with the solution of Young-Laplace equation by FEM. The tangential tensile strength depends on the liquid transition within droplets and the variable mean curvature.

References

[1] K.Murase, H.Shibata, Y.Yasuda and S.Toyama, *Jpn J. Multiphase Flow,* **5**, 322(1991)
[2] A.E.Saez and R.G.Carbonell, *J. Colloid Interface Sci.,* **140**, 408(1990)
[3] D.N.Mazzone, G.I.Tardos and R.Peffer, *Powder Tech.*, **51**, 71(1987)
[4] C.D.Willet, M.J.Adama, S.A.Johnson and J.P.K.Seville, *Langmuir*, **16**, 9405(2000)
[5] M.J.Adama, S.A.Johnson, J.P.K.Seville and C.D.Willet, *Langmuir*, **18**, 6180(2002)
[6] R.K.Niven, *J Petroleum Sci. Eng.*, **52**, 1(2006)
[7] O.Pitois, P.Moucheront and X.Chateau, *Eur. Phys. J.B*, **23**, 79(2001)
[8] S.J.R.Simons, J.P.K. Seville and M.J. Adams, *Chem. Eng. Sci.*, **49**, 2331(1994)
[9] C.D.Willet, M.J.Adama,S.A.Johnson and J.P.K.Seville, , *Powder Tech.*, **130**, 63(2003)
[10] S.M.Iveson, J.A.Beathe and N.W.Page, *Powder Tech.*, **127,** 149(2002)
[11] R.A.Fisher, *J. Agri. Sci.*, 16,492(1926)
[12] F.M.Orr, L.E.Scriven and A.P.Rivas, *J. Fluid Mech.*, **67**, 723(1975)
[13] O.Pitois, P.Moucheront and X.Chateau, *J. Colloid Interface Sci.*, **231**, 26(2000)
[14] U.El Shamy and T.Groeger, *Int. J. Numer. Anal. Meth. Geomeh.*, **32**, 1763(2008)
[15] F.M.Orr, R.A.Brown and L.E.Scriven, *J. Colloid Interface Sci.,* **60**, 137(1977)
[16] K.Murase, T.Mochida, Y.Sagawa and H.Sugama, *Adv. PowderTech.,* **19**, 349(2008)
[17] C.D.Willet, M.J.Adama,S.A.Johnson and J.P.K.Seville, *Handbook of powder technology*, **11**, chap.28, ed. A.D.Salman, M.J.Hounslow and J.P.K.Seville,Elsevier (2007)

NUMERICAL SIMULATION OF WIRE-PLATE ELECTROSTATIC PRECIPITATOR - EFFECT OF PARTICLE CONCENTRATION

Ding Yang[1], Baoyu Guo[2], Jun Guo[1], Aibing Yu[2]*

[1] *Fujian Longking Co. Ltd, Longyan, Fujian, 364000, China*

[2] *Department of Chemical Engineering, Monash University, Clayton VIC 3800, Australia*

Introduction

Electrostatic Precipitators (ESP) have been widely used to separate fine particles from large scale industrial exhaust gases, such as coal fired power plants. In-depth knowledge and careful design optimization are critical to meeting the increasingly strict emission standard in terms of $PM_{2.5}$ emission limit, e.g., 5-10 mg/Nm^3. Numerical modelling has been employed as a powerful tool in this area. Complicated interactions among electric field, gas flow and particle transport are summarized by Adamiak [1]. To model such a system, various levels of simplifications have been made in published papers.

Among them, Lagrangian particle tracking technique is a widely accepted approach. The mass concentration of ash particles in coal-fire flue gas is typically 10-50 g/Nm^3, corresponding to volume fraction in the order of 10^{-5}. Such a small value is often regarded as sufficiently dilute particle concentration in numerical modeling of multiphase flows. For the case of ESP, complicated interactions among the e-field, gas flow and charged particles are involved, even if the deposited dust layer is ignored. Full coupling between the continuous gas phase and discrete particle phase is computationally expensive to achieve a converged steady solution. To avoid such a high cost, the model for ESP was simplified as so called one-way coupling in many previous works, so that the electric field is solved without the particle phase (Guo et al. [2,3]; Farnoosh et al. [4]). This simplification may be acceptable in the end stage of ESP, but will introduce significant error in the early collection stage. Limited studies reported in the literature have considered the particle charges as a source term in the Poisson's equation of electrostatics. For example, Choi and Fletcher [5] predicted a reduced current for the particle-loaded case (d_p = 1-5μm) compared with the particle-free case, but the corona ion charge density somehow shows an increasing trend along the stream. When fine dust particles are treated, the large specific surface area of the particle phase, combined with reduced collection efficiency, implies high space charge density in the collection region. Its effect was understood as a partial suppression of the corona current [6]. As was experimentally proved by Podlinski et al. [7], space charge of charged particles can contribute to the EHD flow pattern and particle transport. In present work, particle space charge is considered through two-way coupling and interesting results are generated in terms of ESP performance as particle loading changes.

Integrated ESP Modelling

Mathematical Equation

Coupled electric field, gas-particle flow and particle charging processes are simulated using

Eulerian-Lagrangian method in Computational Fluid Dynamics (CFD). Major governing equations of vector format are listed below for completeness. For electric field, Poisson equation of potential

$$\nabla^2 V = -(\rho_{ion} + \rho_c)/\varepsilon_0 \quad \mathbf{E} = -\nabla V \tag{1}$$

Current continuity equation

$$\nabla \bullet ((b\mathbf{E} + \mathbf{U}_f)\,\rho_{ion} - D\nabla\rho_{ion}) = S_c \tag{2}$$

where ρ_{ion} is corona charge density, b the ion mobility, D is ion diffusivity, ρ_c is particle space charge density, evaluated by averaging the charges of particles that pass through a control volume (cell),

$$\rho_c = \frac{1}{V_{cell}} \sum_{n=1}^{N} \dot{n}_p q_p \Delta t \tag{3}$$

Where $\dot{n}_p$ is the number flow rate of a parcel of particles tracked, Δt is the residence time in the cell. The source term, S_c, in Equation (2) is transfer rate of ion charges to particles, which is not considered at the moment, i.e., $S_c = 0$. For the gas flow, the momentum equation is,

$$\begin{aligned}\nabla \bullet (\rho_g \mathbf{U}_g \otimes \mathbf{U}_g) = \nabla(p + \tfrac{2}{3}\rho_g k) + \nabla \bullet [(\mu + \mu_t)(\nabla\mathbf{U}_g + (\nabla\mathbf{U}_g)^{\mathrm{T}})] + \rho_{ion}\mathbf{E} \\ -\frac{3\pi n_p d_p \mu}{\mathrm{C_C}}(1 + \mathrm{Re}_p^{0.687})(\mathbf{U}_g - \mathbf{U}_p)\end{aligned} \tag{4}$$

For the particle flow, the force balance of a single particle is

$$m_p \frac{d\mathbf{U}_{\mathrm{P}}}{dt} = \frac{3\pi n_p d_p \mu}{\mathrm{C_C}}(1 + \mathrm{Re}_p^{0.687})(\mathbf{U}_g - \mathbf{U}_p) + \tfrac{1}{6}\pi d_p^3 \mathbf{g}(\rho_p - \rho_g) + n_e q_e \mathbf{E} + \mathbf{F}_{dis} \tag{5}$$

k-ω SST turbulence model is chosen for gas flow, and the dimensionless Lawless charging rate equation is used for particle charging [8].

Model Conditions

An integrated ESP model developed comprises two parallel modules or submodels, namely, electric field model and coupled gas-particle flow model. One-way coupling between the two modules is used earlier is upgraded to two-way in the present work. The electrostatic field caused by charged particles in suspension is considered in the solution of electric field. The two submodels are coupled through a small under-relaxation factor, in views of strong interaction between the electric field and particles flow, and the stochastic nature of the discrete particle phase. In this way all the volume averaged quantities related to the particle phase (such as the particle space charge density) are smoothed out over time in the simulation.

A single channel, 4-wire and flat plate configuration is considered. The dimensions are 3.5mm in wire diameter, 0.24m for wire-wire spacing, 0.2m for wire-plate spacing. The plate length is 1.2m. The simulation domain is defined as 1.74m long by 0.4m wide box with the 4 holes inside. The grid is smoothly refined at places where higher gradient in electric quantities is expected, such as the proximity of the wires and middle of the adjacent wires. Other base case conditions are as follows.

- Applied voltage on the wire electrode: Va = -60kV. The charge density on the wire surface is determined using an injection law combined with Peek equation (Guo et al.[9]).
- Gas temperature: 120°C. As the inlet boundary condition for the gas phase, the velocity is set as 1 Nm/s; relative turbulence intensity 3.7%, turbulent length scale $0.3D_h = 0.24$m.
- Dust particle size is mono-dispersed at d_p=2.5 μm. Particle density 2000 kg/m^3; ash dielectric constant of κ_p=6.6, Particle loading at the inlet, $(\rho_p)_{in}$, is varied with default value of 10 g/Nm^3.

Typical Results

Electric Field without Particles

When the inlet/outlet boundaries are moved sufficient long distance away from the electrodes in the simulation domain (referred to as "natural boundary condition"), the electric field around the boundary wires show an asymmetric distribution pattern that differ from that around the inner wires. Electric field-lines/current can go out of the limit of the collection region (Figure 1). Electric field intensity on the collection plate is of interest as it is considered a factor to determine the upper limit of operation before spark-over [10]. Normally the field intensity on the plate is highest at the point that faces exactly the wire center with the least distance. This maximum field intensity on the plate is 3.80×10^5 V/m. However, when the natural boundary condition is used, the highest value (1.06×10^6 V/m) predicted in field intensity appears at the endpoint of the plate due to its sharpness. This is a likely point that would trigger electric spark-over prematurely. Practically such a sharp edge should be avoided. Generally, the field strength at the middle wires is smaller than that at a boundary wire, due to the counteraction of adjacent wires. Then the corona charge density at the boundary wires should be greater than that at the inner wires (Figure 2), which is similar to the simulation result of Kasdi [11].

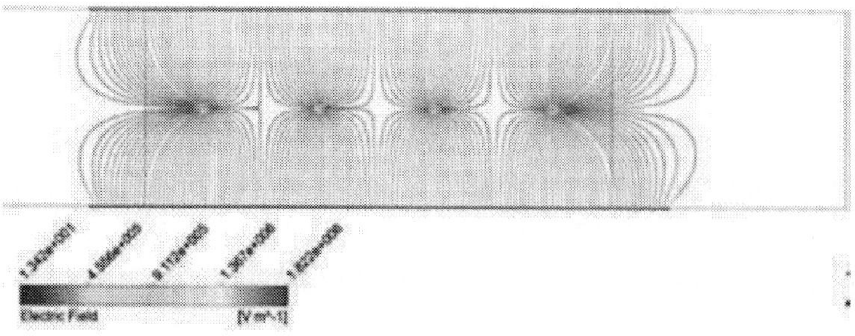

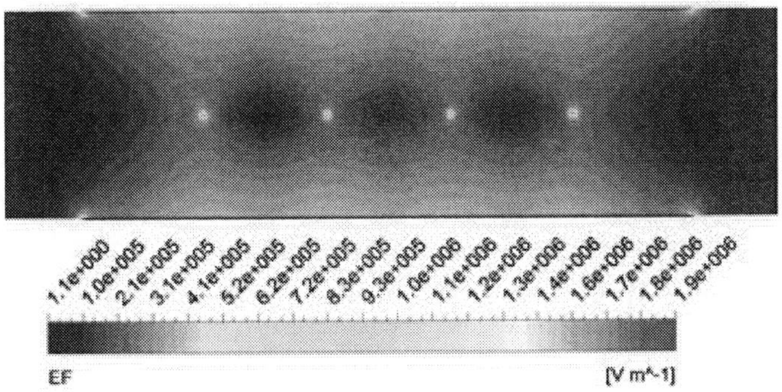

Figure 1. Field streamlines and intensity distribution (no particles).

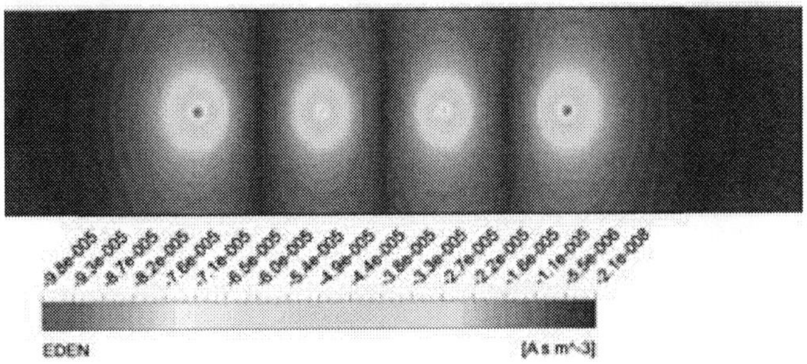

Figure 2. Corona charge density (no particles).

Effect of Particle Space Charge

Particles are randomly injected from the inlet plane with uniform probability in locations and no-slip for velocity (same as the gas velocity). In using random number generator, it is ensured that the injection points are not repeated for every coupling cycle. Figure 3 shows sampled particle trajectories with color scaled to particle charge. Figure 4 shows a pattern of particle deposition rate on the plate. It is seen that the end points of the plate, particularly the downstream one, are special spots to accumulate more dust.

Figure 3. Particle trajectories with colour scaled to particle charge.

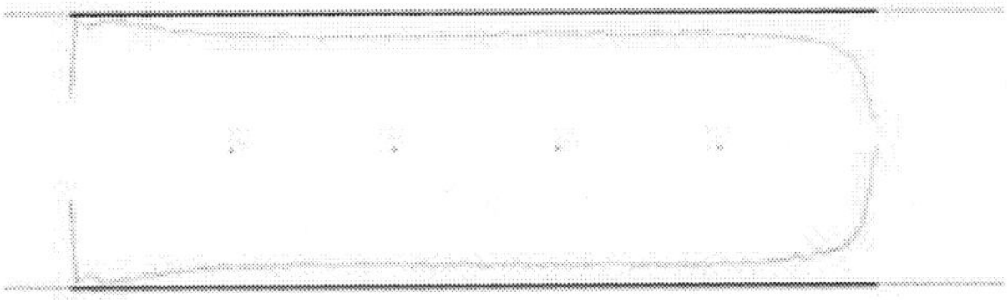

Figure 4. Indicative pattern of particle deposition rate on the plate.

Figure 5 shows the averaged space density of charges carried by particles. Highest space charge density appears near the first wire electrode, where electric current density is highest. It diffuses gradually from the centre line with particles drifting toward the plate while moving forward. Due to the radiating nature of the field-lines around each wire, the lateral component of Coulomb's force in the region close to the centre line may not be as large as that off the centre line. Not collected charged particles of negative polarity are conveyed towards the outlet, lowering the potential downstream outside the collection region (Figure 6), causing distortion of the overall electric field distribution.

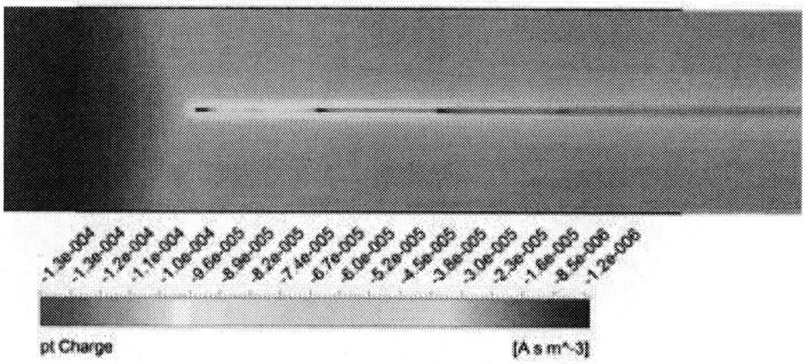

Figure 5. Distribution of space charge density of particles.

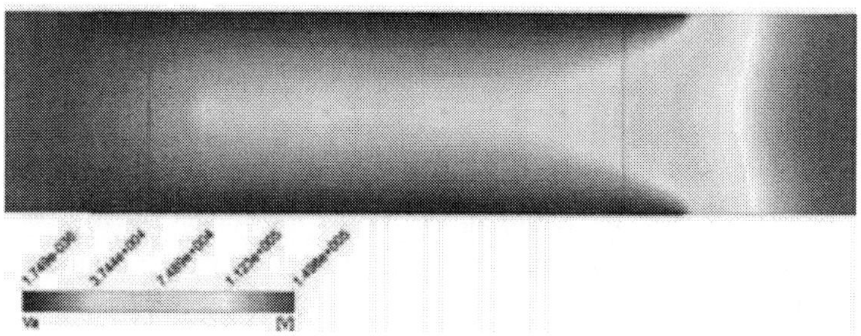

Figure 6. Distribution of electric potential (in absolute value).

Effect of Particle Loading

On Electric Current

Figure 7(a) shows the mean corona current, based on unit length of wire. As can be seen, the corona current drops quickly as particle loading increases, although a fixed voltage is applied. This is caused by the secondary electrostatic field created by the space charges of particles that cancels out the primary electric field at the wire surface and subsequently suppressed the corona generation.

On Particle Collection Efficiency

Predicted particle collection efficiency, as particle loading changes, is given in Figure 7(b). The collection efficiency, increases initially with particle loading (<3 g/m^3), peaks at particle loading of 4-5 g/m^3 and then decreases as particle loading increases further. When ignoring particle space charge, the electric field would be independent of particle loading. This is the case for very low particle loading, where the fractional collection efficiency is constant, or in other words, the quantity of dust collected as well as that penetrated increases proportionally with particle loading.

The observed enhancement of collection by a slight amount of particles is due to the redistribution of electric field so that the near-plate and volume-averaged field intensity happens to increase, which favor the charging of most particles. The asymptotic lower limit of the curve should correspond to the situation when the corona current is so small that turbulent dispersion dominates over the drift force.

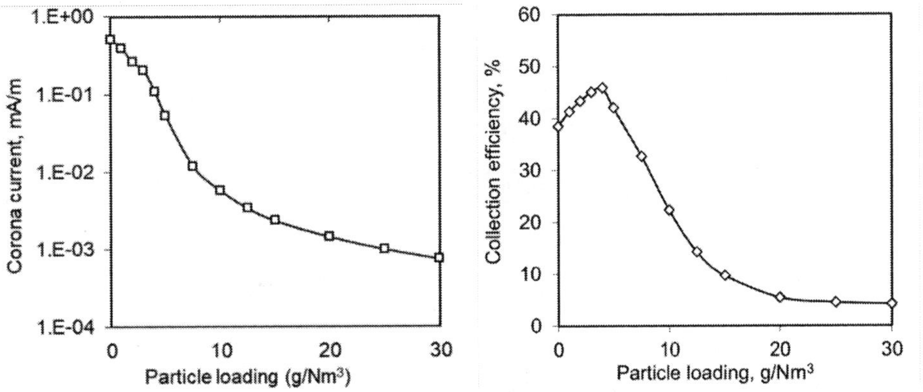

(a) (b)

Figure 7. Mean corona current (a) and collection efficiency (b) as function of particle loading.

On Electric Field Distribution

Figure 8 visualizes the predicted corona charge density and field-lines, which shows how the electric field distribution evolves with particle loading. As the electric field is distorted starting first from the end side, corona density at 3 of the 4 downstream wires is suppressed faster than the 1st wire. When the particle loading is above 7.5 g/m^3, all wires except the first one stop emitting. Meanwhile, reversed field-lines appear around the last wire, and the affected region expands as the loading increases further. Otherwise the field-lines become smoother with increasing loading. Since the field-lines are extended to the outlet, some escaped particles can be pulled back into the collection region, a factor that contributes to significantly greater deposition rate around the end-point of the plate.

Overall, the particle charges affect the particle collection adversely, as they tend to suppress the corona emission or reduce the corona current. Particularly the penetrated particles, which carried significant charges, have prolonged effect as they float along. The corona at the downstream electrodes even extinguish unless the applied voltage is raised individually. When highly charged, more particles precipitate to the plate, reducing the number of particles floating ahead (or less space charge density). The charges carried by these remaining particles exert a negative feedback to the electric field and particle charging, which compromises the collection efficiency.

Electric Quenching

Corona quenching is a common phenomenon in ESP when fine particles are treated. Riebel et al. [12] experimentally studied the corona discharge characteristics of ESP with glycerol aerosol, which found corona onset voltage increases and the corona current significantly decreases after the glycerol steam was injected into the system. Under fixed voltage in the present simulation, corona currents from all the wire electrodes drop with increasing solid loading. The downstream coronas (from the second one) will eventually extinguish at some point. An interesting phenomenon arises when the particle loading increases above 10 g/Nm^3, i.e., the electric streamlines in a small zone around the last wire reverse direction locally. This implies that a charged particle, once falls accidently within the zone, could be trapped and attracted to the same wire, as seen from the particle trajectories in Figure 9. Thus this wire is subject to heavier dust deposition.

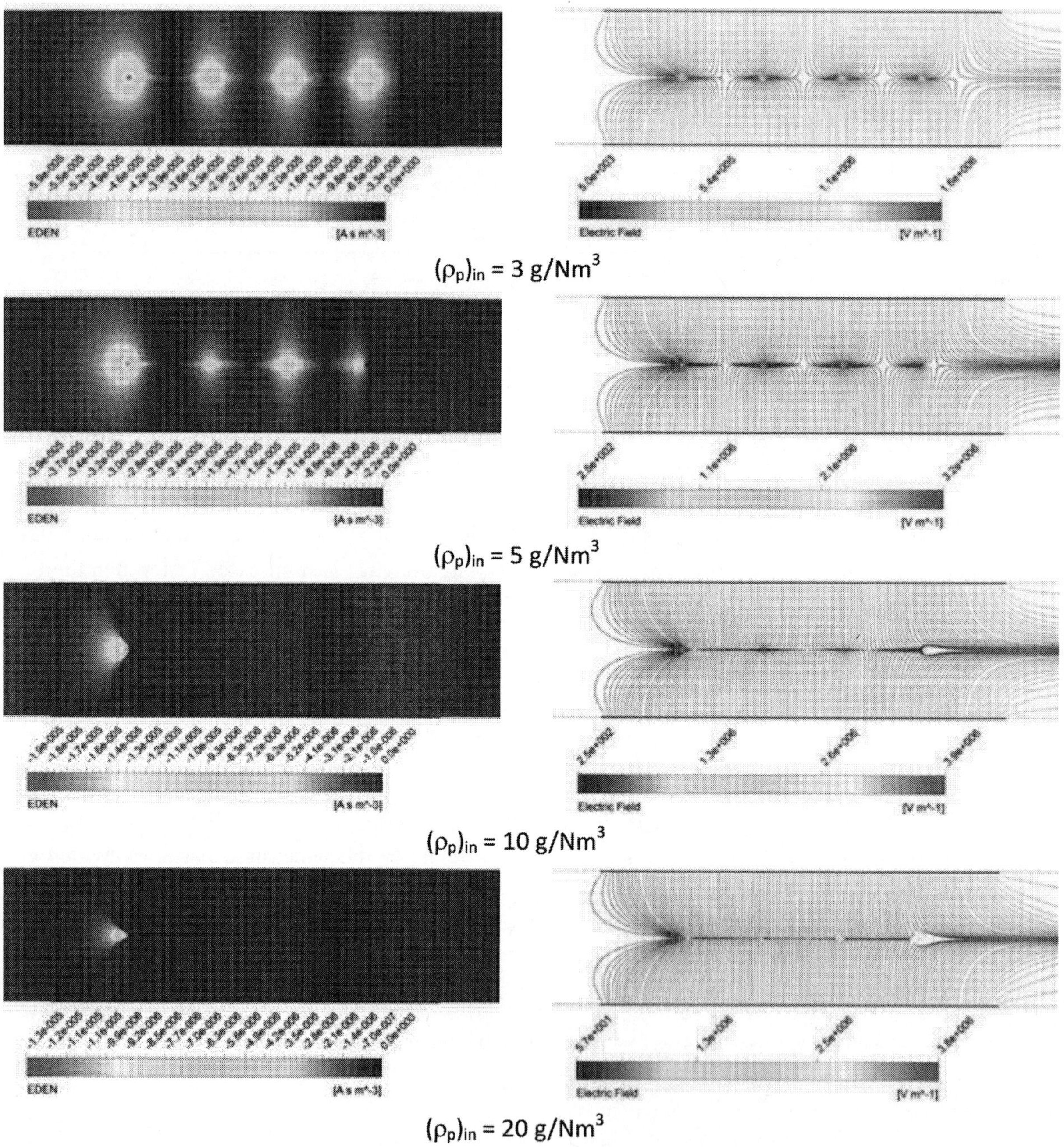

Figure 8. Visualizations of corona charge density and field streamlines for different particle loadings.

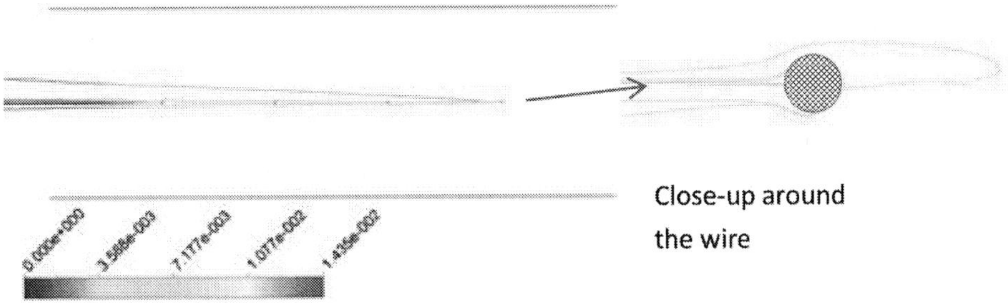

Figure 9. Trajectories of particles falling on the corona wire ($(\rho_p)_{in}$ = 20 g/Nm^3).

Concluding Remarks

CFD based ESP model has been advanced, with an enhanced coupling between electric field and particle flow. So the model can be readily applied to the early collection stage of the ESP, where dust particle loading is too high to neglect. Space charge of particles is considered, and its significance is clarified in a broad range of particle mass concentration in the gas. Dust loading significantly affects the dust collection efficiency for fine particles considered. Maximum collection efficiency exists at a critical particle loading that depends on particle size. However, as particle loading increases, collection efficiency decreases and eventually approaches asymptotically to a lower limit. The phenomenon of electrical quenching is observed, where downstream corona is extinguished or one of the emitters is polluted when the particle loading is high.

References

1. Adamiak K. (2013), "Numerical models in simulating wire-plate electrostatic precipitators: A review", *J. Electrostatics*, 71, pp.673-680.
2. Guo B.Y., Yu A.B. and Guo J. (2014), "Numerical modelling of electrostatic precipitation: effect of gas temperature", *J. Aerosol Science*, 77, pp.102-115.
3. Guo B.Y., Yu A.B. and Guo J. (2015), "Numerical modelling of ESP for design optimization", *Procedia Engineering*, 102, pp.1366-72.
4. N. Farnoosh, K. Adamiak, G.S.P. Castle (2010), "3-D numerical analysis of EHD turbulent flow and mono-disperse charged particle transport and collection in a wire-plate ESP", *J. Electrostatics*, 68, pp.513-522.
5. Choi B.S. and Fletcher C.A.J. (1998), "Turbulent particle dispersion in an electrostatic precipitator," *Applied Mathematical Modelling*, 22, pp.1009-1021.
6. Parker K.R., Applied Electrostatic Precipitation, Blackie Academic & Professional, 1997.
7. Podlinski J., Niewulis A., Mizeraczyk J. and Atten P. (2008), "ESP performance for various dust densities", *J. Electrostatics*, 66, pp.246-253.
8. Lawless P.A. (1996), "Particle charging bounds, symmetry relations, and an analytic charging rate model for the continuum regime", *J. Aerosol Sci.*, 27, pp.191-215.
9. Guo B.Y., Guo J. and Yu A.B. (2014), "Simulation of the electric field in wire-plate type electrostatic precipitators", J. Electrostatics, 72, pp.301-310
10. White H.J. (1974), "Resistivity problems in electrostatic precipitation", *J. Air Pollut. Control Assoc.* 24, pp.313–338.
11. Kasdi A. (2016), "Computation and measurement of corona current density and V–I characteristics in wires-to-plates electrostatic precipitator", *J. Electrostatics*, 81, 1-8.
12. Riebel U., Radtke R. and Loos R. (2002), "An experimental investigation on corona quenching", *J. Electrostatics*, 54, pp.159-165.

Triboelectric Charge of Spherical Glass Particles against Metal Pipeline

HoSu CHOI [1], Teruo SUZUKI [2] and Kwangseok CHOI [3,]*
[1] Christian Academy in Japan, [2] Kasuga Denki Inc.,
[3] National Institute of Occupational Safety and Health, Japan
**choiks@jniosh.go.jp*

Introduction

Insulating powders are used in many ways such as in the food, pharmacy, and chemical industries. Usually, these powders are highly charged when transported, loaded, dried, and mixed, due to repeated contact and separation between powder particles and the equipment. This results in the powders acquiring electrostatic charges. If not dissipated quickly, electrostatic charge buildup can cause inter-particle cohesion and adhesion to operating equipment, causing a production efficiency decrease. Incendiary electrostatic discharges such as spark discharges, cone discharges, and propagation brush discharges can also take place during industrial operations and may lead to fires and even explosions. Riedewald reported on an industrial fire in a pharmaceutical powder transfer operation [1]. A large number of studies involving electrostatic charge generation on powders, as well as electrostatic discharges and hazards during operation, have been reported. For example, Kaialy reviewed the main factors affecting the electrostatic charging of pharmaceuticals such as particle resistivity, size, shape, and surface roughness [2]. Gibson also reviewed industrial hazards related to electrostatic charges and discharges [3]. Our previous studies [4–6] investigated the electrostatic charging of polymer powder in mixing processes and the ignitability of aluminous, magnesium and polymer powders due to electrostatic discharges involving the effect of the ratio of N2 in the air on the ignitability.

In order to easily measure and confirm the electrostatic charge level of powders, a novel tribocharging system that utilizes the Faraday cup is presented, and the charging behavior of the sample powders is investigated in this paper.

Experimental Apparatus and method

Figure 1 shows the spiral air tribocharge apparatus presented in this study. The apparatus was divided into four main parts: the air supply part, the powder supply part, the tribocharge part, and powder collection part. The air supply part consisted of a compressor (Hitachi, PO3.7MA) that included an auto dryer (CKD, F3-217199), air controller, electronic valve, and switch. In this study, dried air was used because moisture in the air may have an effect on the electrostatic characteristics of the powder sample. The powder supply and tribocharge parts consisted of a powder port, a spiral air flow device, and a pipeline. The spiral air flow device (photo 1) was attached in front of the sample powder port in order to create spiral air inside the pipeline. The spiral air flow device had 8 holes that were 6 mm in diameter and that opened at a slant (60 degrees), which is very different from conventional models [7, 8, 9]. The straight pipeline was made up of stainless steel metal and was 1300 mm in length and 35 mm in diameter. In order to prevent external noise, the overall powder supply and tribocharge parts were shielded. The powder collection part was made up of a cyclone-type Faraday cup, unlike conventional models [7, 8, 9]. The cyclone-type Faraday cup had a separate metal container (30 mm in diameter) for easy measurement of the amount of collected powders. The other devices included two electrometers (Kethley,

6517A), a data logger (Graphtec, GL900), an electrostatic ionizer (Kasuga Denki Inc., KD-510W), an electrical balance (Shimazu, BL220H), and a timer.

The typical operating procedure for the tribocharge test is summarized as follows:
(1) Adjust the air pressure using the air pressure valve.
(2) Load 1 g of the sample powder into the powder port.
(3) Eliminate the initial charge of the sample powders using an electrostatic ionizer.
(4) Reset the electrometers (Q mode) connected to the pipeline and cyclone.
(5) Switch the air flow on, and run the sample powder through the pipeline where friction occurs.
(6) Measure the indication value when it stabilizes.
(7) Measure the amount of powder collected in the Faraday cup.
(8) Clean out the pipeline and cyclone part using an air spray in preparation for the next test.

The charge to mass ratio, q (nC/g), of the sample powder was obtained by dividing the charge, Q (nC), by mass, m (g), of the sample powder loaded into the Faraday cup:

$$q = \frac{Q}{m} \quad \text{..} \quad (1)$$

This experiment is mainly to determine how much the charge to mass ratio, q (nC/g), of the sample powders is affected by the size of the powder and the amount of air pressure. The cyclone-type Faraday cup and the pipeline were suspended electrically with insulators. All experiments were conducted in ambient conditions (air temperature: 22 ℃; relative humidity: 55 %).

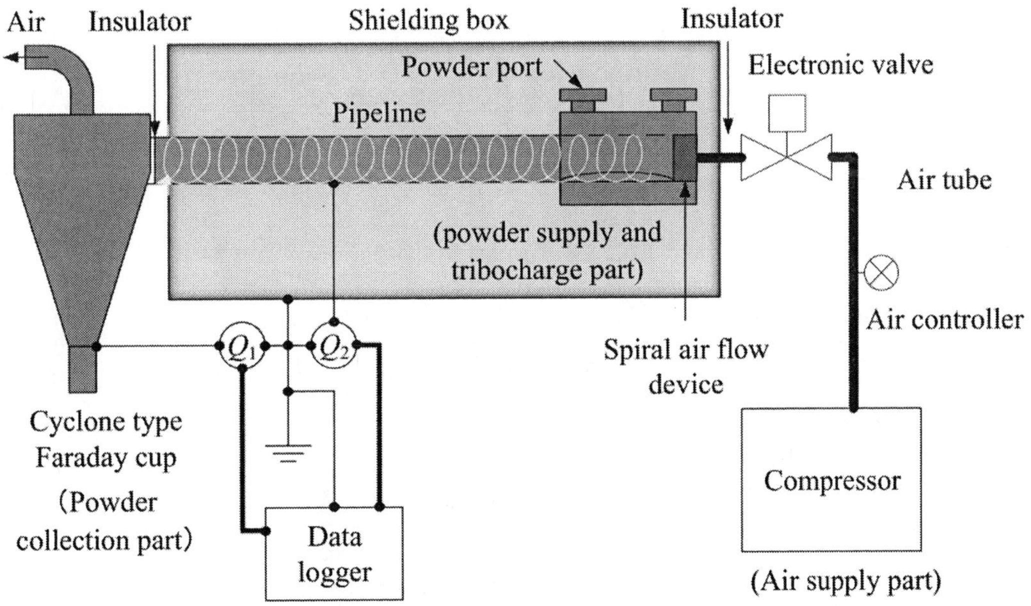

Fig. 1 Schematic of the spiral air tribocharge apparatus used in this study.

Photo 1 Structure of the powder port and spiral air flow device.

Sample powders

Glass beads are extremely good light reflectors and thus are widely used as the main component of the coating material for light-reflecting adhesive tape, background paint for traffic control sign boards, and paint for road signs. Four different grades of commercial glass beads were used in this experiment, as shown in Table 1. The particle size was determined using a laser diffraction system (LDA Win 1.21, dry type). Hereafter, they will be called by their abbreviation shown in Table 1. The analysis of the images from microscopy (Neoscope, JCM-500) revealed that the G-4 powder particles were spherical, as shown in photo 2. The other glass bead particles were the same shape as G-4.

Table 1. List of powder specimens used in experiments.

Specimen	Abbreviation	Diameter [μm]		
		D_{10}	D_{50}	D_{90}
Glass beads 1	G-1	22	62	85
Glass beads 2	G-2	93	131	175
Glass beads 3	G-3	130	180	241
Glass beads 4	G-4	162	225	294

D_X: X% volume average.

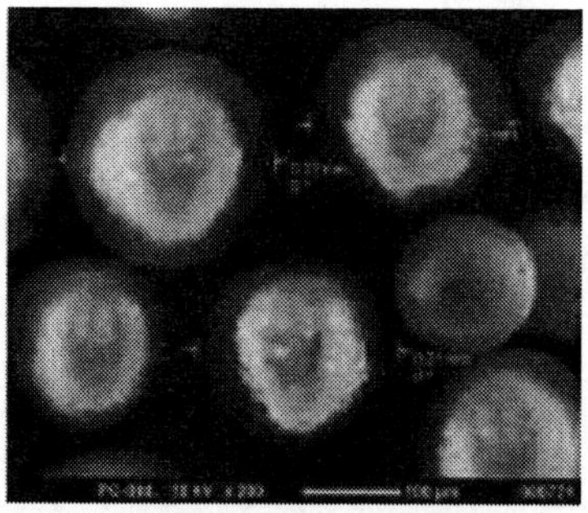

Photo 2 Shape of G-4 powder used in this experiment.

Experimental results and discussion

Figure 2 shows the typical change of charge detected over time when the G-3 particle was used and the air pressure was 0.15 MPa. It should be noted here that when no particle was used and only air was passed through the pipeline, the charge was detected as almost zero. The charge Q1 of the powder was found through the cyclone-type Faraday cup and then was confirmed through the charge Q2 obtained from the pipeline. As the powder particles passed through the pipeline for about 0.82 seconds, as seen in the Fig. 2, the charge clearly increased. Q1 and Q2 had almost the same trend and absolute values. This result confirmed the reliability of the test and showed that there was no disturbance by external noise.

In this study, all of the glass beads (G-1 to G-4) were negatively charged when stainless steel metal was used as the contact surface. The charge polarity of the materials relates to the work function (or ionization potential) of each material. The contact potential difference between two substances is defined by the difference between their fermi levels, which correlate with the work function. In general, materials with higher work functions tend to appropriate electrons from materials with lower work functions. This point will be carried out in our further study.

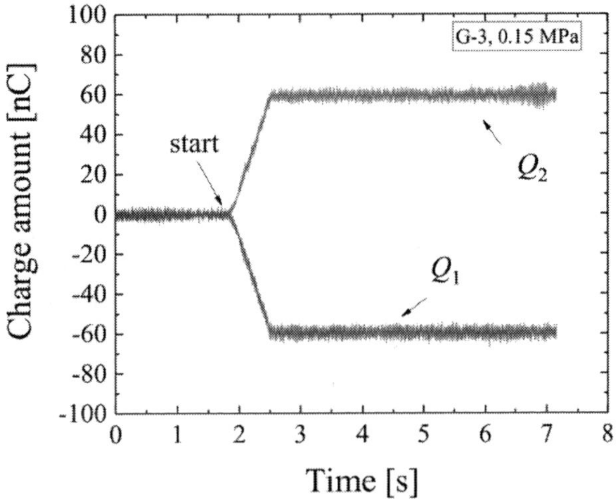

Fig. 2 Change of charge detected over time.

Figure 3 shows the charge to mass ratio of glass beads as a function of applied air. The range of applied air was between 0.1 MPa and 0.3 MPa in this study. When the applied air pressure was below 0.1 MPa, it was most likely not enough to send all of the powder through the pipeline. Each experiment was repeated five times, and the average values were reported. This graph does not contain an error bar, but these were about 10%. The results showed that the charge to mass ratio of both powders increased as the applied air increased. Although not shown in Figs. 2 and 3, the increase in air pressure also causes the slope of the graph to increase significantly. This can be explained by the fact that with increased pressure of applied air, the particles collide against the surface of the pipeline with more strength [9]. In order to investigate in more detail, the velocity of air at approximately 10 mm from the inside of the pipeline wall was measured using an anemometer (Kanomax, Anemomaster-6115). The approximate velocity of air flow was 9.5 m/s at 0.1 MPa, 12 m/s at 0.15 MPa, 15 m/s at 0.2 MPa, 20 m/s at 0.25 MPa, and 22 m/s at 0.3 MPa. It could be clearly seen that as the air pressure increased, the air velocity also increased. Another reason for this may be that because the air was applied in a spiral form, the particles may have collided more frequently against the pipeline when the air flow was increased. This exception will be investigated in detail in the future.

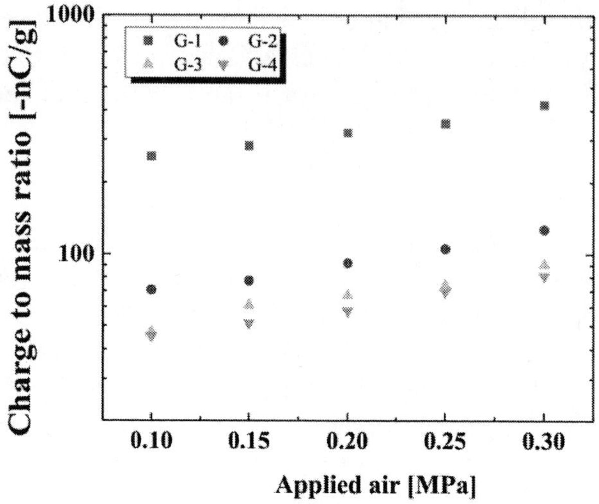

Fig. 3 Charge to mass ratio of glass beads as a function of applied air.

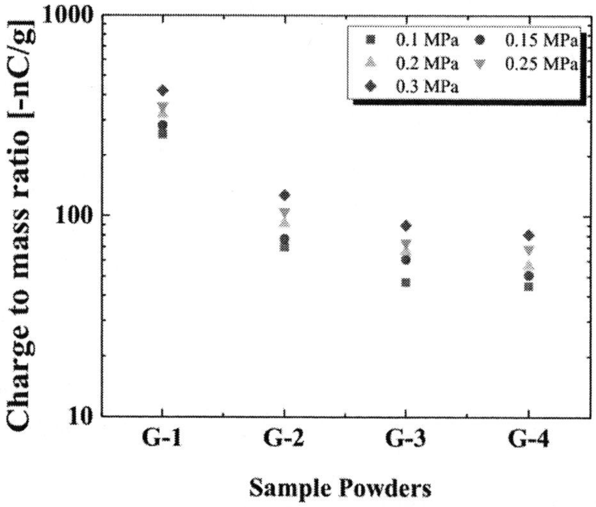

Fig. 4 Charge to mass ratio as a function of different glass bead sizes.

Figure 4 shows the charge to mass ratio as a function of different sizes of glass beads. The size distribution of each glass bead in this study was described in Table 1. The results showed that the charge to mass ratio of the glass beads increased as the powder size decreased. That is to say, smaller particles had a larger q than the larger ones. The total surface area of all of the smaller beads combined is greater than that of the larger ones, which results in a high charge to mass ratio development in the tribocharge. The value range of the q of the glass beads obtained in this study was about 40 nC/g to 100 nC/g. These values agreed with those obtained from the micronizing and pneumatic transfer process [10]. These values were high enough to generate incendiary electrostatic discharges during handling in industrial processes. This suggests that if glass beads are used with solvents in the cleaning process, they must be handled carefully, as incendiary discharges can ignite the solvents even though the glass beads themselves are nonflammable material.

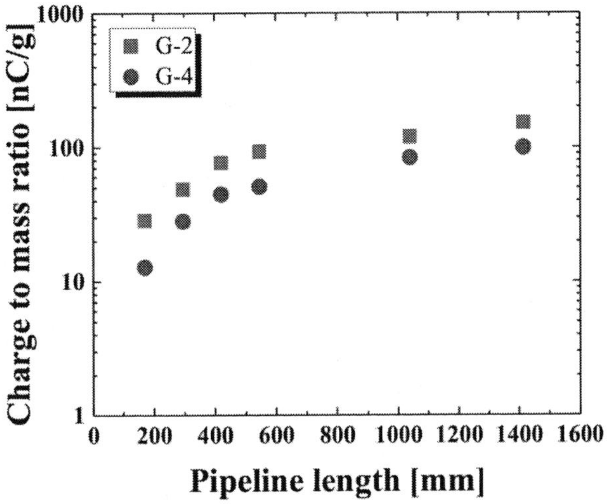

Fig. 5 Charge to mass ratio as a function of pipeline length.

Figure 5 shows the charge to mass ratio as a function of pipeline length. Six different pipeline lengths, *d*, were used: 170, 295, 420, 545, 1040, and 1415 mm. The glass beads used in this experiment were G-2 and G-4, and the air pressure was 0.35 MPa. Since the pipeline length changed with each test, it was difficult to attach the Faraday cup to the end of the pipeline. Therefore, in this study, only the value of q_2 was measured. As seen in Fig. 2, q_2 had the almost same value as q_1 but had an opposite sign. The length of the pipeline was measured starting from beginning of the powder port because the inside structure of the powder port was the same as that of the pipeline. Each experiment was repeated five times, and the average values were reported. The results showed that q_2 increased as the length of the pipeline increased for both sizes of glass beads. This can be explained by the fact that as the length increases, the time the glass beads come in contact with the pipeline also increases. The increase rate at the beginning section of the pipeline was especially large. It can also be seen through Fig. 5 that overall, G-2 had a larger q_2 than G-4 because of the total surface area. This confirms the reliability of the data shown in Fig. 4.

In addition, both beads seemed to reach saturation when the pipeline length was 1415 mm. It was compared to the maximum charge to mass ratio, q_c [nC/g] calculated from the well-known equation shown below [11]:

$$q_c = |\sigma_{max}|/(D \cdot \rho), \quad \text{-------------------------(2)}$$

where σ_{max} [kg/m^3] is the maximum surface charge density (27 μC/m^2); D [μm] is the particle diameter; and ρ is the particle density (1.5×10^3 kg/m^3).

In this study, the particle sizes of G-2 and G-4 were 131 μm and 225 μm, respectively, as shown in Table 1. The calculated maximum values q_c and experimental maximum values q_t are compared as shown in Table 2. The calculated and experimental values for both G-2 and G-4 agreed well with each other. Therefore, it can be seen that both G-2 and G-4 reached saturation.

Table 2. Calculated maximum values and experimental maximum values of the charge to mass ratio.

	q_c [nC/g]	q_t [nC/g]
G-2	137	151
G-4	80	99

Conclusion

In this paper, a novel tribocharging system that utilizes the Faraday cup was presented and the electrostatic charge behavior of glass beads was experimentally investigated. The results were as follows:
1) For all of the glass beads, the charge clearly increased as the glass bead particles passed through the pipeline. The charge amount of the glass beads and the charge amount obtained from the pipeline had almost the same trend and absolute values. This result confirmed the reliability of the test and showed that there was no disturbance by external noise.
2) All of the glass beads were negatively charged when stainless steel metal was used as the contact surface. The value range of the charge to mass ratio of the glass beads obtained in this study was about 40 nC/g to 100 nC/g.
3) The charge amount of the glass beads increased as the applied air increased. It could also be seen that as the air pressure increased, the air velocity also increased.
4) The charge amount of the glass beads increased as the powder size decreased due to the total surface area of the beads.
5) The charge amount of the glass beads (G-2, G-4) increased as the length of the pipeline. The increase rate at the beginning section of the pipeline was especially large.

Acknowledgments
The authors gratefully acknowledge Mr. Naoto Nogera for his assistance during the test.

References

1. F. Riedewald (2012): "A fire in a secondary pharmaceutical powder transfer operation," *Process safety progress*, 31, 4, pp. 390-392.
2. W. Kaialy W, "A review of factors affecting electrostatic charging of pharmaceuticals and adhesive mixtures for inhalation," *Int.J.Pharm.* Vol.503, pp.262-276.
3. N. Gibson (1997), "Static electricity — an industrial hazard under control," *J. Electrostatics,* Vol. 40-41, pp. 21-30
4. K. Choi, K. Choi, K. Nishimura (2015), "Experimental study on the influence of the nitrogen concentration in the air on the minimum ignition energies of combustible powders due to electrostatic discharges," *J. of Loss Prevention in the Process Industries.* Vol.34.pp.163-166.
5. K.S. Choi, H.Sakasai, and K. Nishimura (2015), "Experimental study on ignitability of pure aluminum powders due to electrostatic discharges and Nitrogen's effect," *J. of Loss Prevention in the Process Industries.* Vol.35.pp.232-235.

6. K.S. Choi, H. Sakasai, and K. Nishimura (2015), "Minimum ignition energies of pure magnesium powders due to electrostatic discharges and Nitrogen's effect," *J. of Loss Prevention in the Process Industries.* Vol.41.pp.144-146.
7. N. Masui, and Y. Murata (1981), "charging tendency of the flowing powder through a metal tube (II), " *The five Annual meeting of Institute of Electrostatic Japan,* pp. 244-247.
8. S. Watano, S. Saito, and T. Suzuki (2002), "Feedback control of powder electrostatic charge in a pneumatic conveying process," *J. Soc. Powder,Japan*, Vol. 39, pp. 496-502.
9. P. Boschung, M. Glor (1980), "Methods for investigating the electrostatic behavior of powders, J. of elelctrostatic," Vol. 8, pp. 205-219.
10. IEC50404(2004), "Electrostatics- Code of practice for the avoidance of hazards due to static electricity," p. 66.
11. M.Glor (1988), "Electrostatic hazards in powder handling," *Research studies press LTD.*, p.26.

509266 Spray-Drying of a Layered Double Hydroxide Nanosuspension

Abstract:

Over the past few years, layered double hydroxide (LDH) has attracted particular interests in both academia and industry. LDH is 2D anionic clay of brucite-like layered structure with various organic or inorganic anions located between layers. The diverse composition of LDH, and its associated physical and chemical properties open up new possibilities in various applications. LDHs are successfully applied as precursors in synthesis of supported catalysts, sorbents, carriers delivering bioactive molecules, matrices containing active pharmaceutical ingredients for control release, electrochemical biosensors, etc. LDHs are mainly synthesized in a form of nanosuspensions. The application of nano-sized sorbents or catalysts in the liquid-phase processes requires a special technique to separate them from the final product. An uncontrollable agglomeration of solid nanoparticles makes their industrial application impossible in a dry phase. Moreover, the toxic effects of nanoparticles on human health and ecosystem have been discussed in the literature.

Utilization of larger micron-size agglomerates of nanoparticles is one of the alternative ways to safely handle nanoparticles in the dry phase on the industrial scale. Spray drying is one of the well-known techniques that have been widely applied for production of microspheres from nano-sized particles. Using this technique, it is possible to control the size distributions as well as the morphology of spray-dried agglomerates. However, there are only a limited number of publications related to spray drying of LDH nanosuspensions. Moreover, spray-dried LDH agglomerates of rather wide size distribution, low surface area and irregular shape were reported in the literature.

In this study, we explore the effect of process parameters, such as spraying air flow rate, drying air temperature, slurry concentration and slurry feed rate on the size distribution, morphology, specific surface area and pore volume of LDH agglomerates fabricated by spray drying of nanosuspensions.

The round-shaped agglomerates with a diameter of more than 2 and less than 15 micrometers were obtained by spray drying of suspensions of nanoparticles with median diameter of 171 nm without addition of dispersant and binder. Our results confirmed that the spraying air flow rate has the dominant effect on the agglomerate median size and the width of agglomerate size distribution as well as on the agglomerate specific surface area. The small agglomerates with narrow size distribution were collected during drying at the high flow rate of spraying air. Our results also indicate that lowering the temperature of drying air and using the dilute nanosuspensions promote generation of small agglomerates.

The specific surface area of spray-dried LDH agglomerates is in the range of 49 to 101 m^2/g. The agglomerates of high specific surface area were obtained at low flow rate of spraying air, high slurry feed rate of concentrated slurry and using low temperature drying air. These results also confirm that drying at low drying rate yields to formation of agglomerates of high surface area. Similar trends were observed in terms of the variation of total pore volume with drying process parameters.

Thus, by adjusting the spray drying conditions, it is possible to produce agglomerates of LDH nanoparticles satisfying the product requirements.

513829 Impact of Spheroidization of UO_2 Powders on the Filling of Press Molds

Abstract:

In industry, the use of powders is very common. This underlines the interest in understanding the flow of powders in process lines (hopper, silos, transportâ€¦) and finding the link between powder physical characteristics and its rheological properties. This study deals with the flowing properties of a UO2 powder in a powder metallurgy process for nuclear fuel fabrication. A reproducible filling of the press molds is required to avoid mechanical grinding of the sintered pellets.

A powder is characterized by its particle size distribution but also by the morphology of particles. Morphology has a major impact on the cohesion and then on the flowability of powders [1]. UO2 powders are generally made up agglomerates which are composed of aggregates, themselves formed by crystallites (Figure 1). Agglomerates are primarily governing the powder flowing behavior. Their characteristics sometimes do not allow them to meet the expectations of the process, especially for the filling of the press molds.

For this reason, in a powder manufacturing process, a spheroidization operation (Figure 2) could precede the shaping of pellets to improve the filling properties.

In this study we observe the influence of particles morphology on the rheological properties of UO2 powders. Whereas there is no standard and universal method to characterize the flow of powders [2], there is direct (shear cells) and indirect methods (dynamic flowability measurements like permeability). We use both methods: the indirect methods based on qualitative measures of powder cohesion, the direct methods to determine the powder cohesion for a specific consolidation state through the Freeman shear cell [3]. While a shear cell measurement permits to determine the powder flow in quasi-static system, a permeability measurement shows the flow index in dynamic system. We identify the link between theses measurements and the quality of the filling.

References:
[1] Xiaowei Fu and others, «Effect of Particle Shape and Size on Flow Properties of Lactose Powders», Particuology, Advances in Characterization and Modeling of Particulate Processes, Apr. 2012.
[2] P. Juliano, B. Muhunthan, and G. V. Barbosa-Cánovas, « Flow and shear descriptors of preconsolidated food powders », J. Food Eng., vol. 72, no 2, p. 157-166, Jan. 2006.
[3] J. Schwedes and D. Schulze, « Measurement of flow properties of bulk solids », Powder Technol., vol. 61, no 1, p. 59-68, Apr. 1990.

508203 Modelling a Twin Screw Granulator Using the Discrete Element Method

Abstract:

Wet granulation is a process used to create larger stable agglomerates (granules) from fine powders. This has many desirable outcomes such as improving flowability, compactibility, and homogeneity. Granulation is commonly employed in the food, pharmaceutical, detergent, and fertilizer industries, but despite its wide adoption it is often inefficiently operated, with high recycle ratios in continuous processes and high rejection rates in batch processes. Wet granulation is the most common type of granulation and, in pharmaceutical applications, it is a critical step in tablet manufacturing that affects the uniformity and compactibility of the final dosage form.

Traditionally batch granulation was the favoured granulation process method in the pharmaceutical industry due to the challenges faced from adopting a continuous approach, such as cost or inability to monitor product quality reliably. Since the introduction of Quality by Design (QbD) and Process Analytical Technology (PAT) by the FDA in 2003 there has been a move towards new manufacturing techniques such as continuous processing.

Although tremendous efforts have been made to gain scientific insight into the granulating process, a fundamental understanding of wet granulation is still lacking due to the complexity of the mechanisms involved. With twin screw granulation becoming a popularly employed method of wet granulation, an in-depth understanding of particle enlargement in the granulating process is necessary in order to improve the quality of the final product without the need for large-scale Design-of-Experiment studies.

Despite the extensive experimental research carried out for twin screw granulators in recent years, there has been little computational work carried out. This paper employs the Discrete Element Method (DEM) to study a 25 mm diameter, GEA ConsiGmaâ„¢ 1 twin screw granulator with a typical 60Â° forward configuration for kneading elements. The DEM simulations were conducted using the commercial code EDEM with a DEM contact model developed for cohesive solids implemented through an API. The contact model is based on an elasto-plastic contact with adhesion and uses hysteretic loading and unloading paths to model the elastic-plastic contact deformation. In these simulations, the adhesion is used to capture the effect of the binder liquid without the complication of modelling the liquid directly. The adhesion parameter is a function of the plastic contact overlap. The model has previously been shown to be able to predict the stress-history-dependent behaviour depicted by a flow function of the material. In this study the effect of particle size, gap tolerance and cohesion on the residence time distribution and local solid fraction at different elements in the granulator is assessed. The level of cohesion used is related back to uniaxial test results which can be used to measure the flowability of a cohesive granular solid.

508216 Hybrid Multiscale Modelling of a Twin Screw Granulator

Abstract:

Wet granulation is a process used to create larger stable agglomerates (granules) from fine powders. This has many desirable outcomes such as improving flowability, compactibility, and homogeneity. Granulation is commonly employed in the food, pharmaceutical, detergent, and fertilizer industries, but despite its wide adoption it is often inefficiently operated, with high recycle ratios in continuous processes and high rejection rates in batch processes. Wet granulation is the most common type of granulation and, in pharmaceutical applications, it is a critical step in tablet manufacturing that affects the uniformity and compactibility of the final dosage form.

Modelling processes such as wet granulation is not straightforward, but may be achieved using techniques such as Population Balance Modelling (PBM) and Discrete Element Method (DEM). PBM groups particles into different classes based on their properties and uses empirically derived differential equations to describe key processes such as aggregation or breakage. However, although PBM tracks the size and number of particles over time, it cannot account for the distribution of the particles and the effect of this local variability.

By contrast, DEM simulations track individual particles, and as such, phenomena such as segregation or liquid migration can be studied. DEM simulations also consider the effect that the geometry and its dynamics have on the particles' behaviour, which is something that PBM cannot account for directly. However, DEM simulations of fine materials can be computationally expensive and this often results in reduced numbers of particles being used to make the simulation computationally tractable. This is usually achieved through increased, unrealistic particle sizes or the simulation of isolated local zones without considering adjacent zones which may affect the behaviour.

Traditionally, large-scale Design-of-Experiment (DoE) studies would be needed to estimate key rate parameters for use in PBM modelling of granulation and this can be both costly and time-consuming. DEM simulations provide an alternative method to estimate these parameters. By coupling the two simulation methods, one can utilise the key benefits of each method to better simulate complex processes.

This paper presents a hybrid approach to modelling twin screw granulation via DEM–PBM coupled simulations to study a 25 mm diameter, GEA ConsiGma™ 1 twin screw granulator with a typical 60° forward configuration for kneading elements. The DEM simulation was conducted using the commercial code EDEM with a DEM contact model developed for cohesive solids implemented through an API. The PBM simulation is carried out with the commercial code gFORMULATE which is built on top on the advanced modelling platform gPROMS. A bi-directional coupling is implemented with information shared between the two simulation tools at specified intervals. Preliminary results for this coupled approach are presented and discussed.

509577 Blocking Rules for Discharging Granular Materials from a Flat Bottom Hopper

Abstract:

Hoppers and silos are widely used in storing granular materials in various industries, such as agricultural, chemical, food and pharmaceutical industries. Three key aspects need to be considered when hoppers are used to store powders and grains: 1) stress analysis; 2) hopper blockage ; 3) mass flow rate. A hopper must be designed strong enough to sustain the stress generated from the bulk materials during storage and discharge. On the other hand, a hopper must be designed properly so that the stored bulk materials can be discharged, i.e. without blockage and formation of ratholes and arches. Furthermore, the mass flow rate needs to be predicted and controlled. In this study, a combined experimental and numerical study was performed to determine the blocking rule for discharging granular materials from a flat bottom hopper. Granular materials of various sizes were considered and the minimal orifice size through which the bulk materials can be discharged was determined experimentally using a Flodex tester and numerically using DEM. The dependency of the minimal

orifice size on particle size was explored. An semi-empirical model was also developed. Using this model, the critical orifice size at which the flow will stop, i.e. blockage takes place, can be determined from the mass flow rate data. It was found that excellent agreement between the numerical results and experimental measurements was obtained. Both experimental and numerical results showed that, even for spherical particles, layered dead zones could be formed during the discharge from the flat bottom hopper. In addition, mono-layered dead zones were developed when large particles were used. Furthermore, the developed blocking rules were also verified experimentally and numerically.

508310 Characterization of Mesoscopic Structure in Cohesive Powder, Neat or Blended, By X-Ray Computed Tomography and Prediction By the Discrete Element Method

Abstract:

Understanding and predicting structural characteristics of cohesive powder systems is a technical challenge in many powder technology processes and applications. The discrete element method (DEM) [1] is a promising approach to model the structure and flow behavior of cohesive granular systems. Advances in experimental techniques such as X-ray computed tomography (XRCT) have now enabled characterization of fine cohesive powder systems with sub-micron resolution. Combined with standard bulk powder characterization methods (e.g. density), this technique allows for mesoscopic structural characteristics of cohesive powder systems to be compared with DEM simulation. In this study, DEM and XRCT are used to characterize the mesoscale structure, under controlled consolidation conditions, of two moderately cohesive 'model' (roughly spherical) powders with distinct material properties. The DEM simulations are based on a constitutive elasto-plastic model [2] that makes use of the JKR theory [3] to account for cohesive interparticle forces. Density and void structure obtained from DEM simulation are compared to characterization results obtained from XRCT image analysis. XRCT and DEM simulation approach is applied to examine structures of binary mixtures of the two model powders. The ability of DEM simulation and the contact model to accurately predict structural characteristics of single component and binary granular systems is discussed.

References

[1] P. A. Cundall and Otto D.L. Strack. A discrete numerical model for granular assemblies. Geotechnique, 29(1):47–65, 1979.

[2] C. Thornton and Z. Ning. A theoretical model for the stick/bounce behavior of adhesive, elastic-plastic spheres. Powder technology, 99(2):154– 162, 1998.

[3] K.L. Johnson, K. Kendall, and A.D. Roberts. Surface energy and the contact of elastic solids. In Proceedings of the Royal Society of London A: Mathematical, Physical and Engineering Sciences, volume 324, pages 301–313. The Royal Society, 1971.

511638 Numerical Research of Hydrodynamics in Gas-Solid Micro Fluidized Beds

Abstract:

Due to the advantages of less energy consumption, more safety and good operability, the miniaturization of fluidized bed, or named micro fluidized bed, has been extensively studied by a lot of researchers. The Micro Fluidized Bed Reaction Analyzer (MFBRA) was adopted by the Institute of Process Engineering (IPE), Chinese Academy of Sciences (CAS), to study the gas-solid reaction kinetics under isothermal differential condition[1, 2]. In the MFBRA, the hydrodynamics of micro fluidized beds play a vital role on the heat and mass transfer and which is a key to the accurate measurement of gas-solid reaction kinetics. In order to optimize and design the MFBRA, numerous experiments have been conducted to study the hydrodynamics of micro fluidized beds[3, 4]. Unfortunately, those works mostly focused on the bed pressure drop and minimum fluidization velocity, nevertheless, the detail information, such as local solid volume fraction and particle velocity profile in micro fluidized beds are reported rarely. It is because the existing experimental techniques, such as, optical fiber and capacitance probes are hard to measure those local parameters without disturbing the flow field.

The objective of this work is to study the hydrodynamics of micro fluidized beds and to investigate the isothermal characteristics during kinetics testing of graphite powder combustion in MFBRA. In this study, Discrete Particle Model (DPM) and Two-Fluid Model (TFM) are selected to investigate micro fluidized beds, respectively. The DPM is carried out by the commercial software EDEM-Fluent Coupling simulations. And for TFM, we used the Eulerian multiphase model in the ANSYS Fluent. From our numerical simulations, the detailed hydrodynamic characteristics of micro fluidized beds are explored systematically by comparing different column sizes and superficial gas velocity. Meanwhile, it is found that both DPM and TFM have the ability to model the MFBRA under practical operation condition with high superficial gas velocity. In the term of isothermal characteristics in micro fluidized beds, the temperature field distribution as well as the heat transfer between heat-carrier and graphite powder are explored by the combination of simulations and experiments.

Reference:

1. Xu, G., *Micro fluidized bed reaction kinetics analyzer. Supported by instru-ment development project of Chinese Academy of Sciences, China, Project numberY2005014. (in Chinese).* 2005.
2. Yu, J., et al., *Isothermal differential characteristics of gasâ€"solid reaction in micro-fluidized bed reactor.* Fuel, 2013. **103**: p. 29-36.
3. Liu, X., G. Xu, and S. Gao, *Micro fluidized beds: Wall effect and operability.* Chemical Engineering Journal, 2008. **137**(2): p. 302-307.
4. Guo, Q.j., Y. Xu, and X. Yue, *Fluidization Characteristics in Micro-Fluidized Beds of Various Inner Diameters.* Chemical Engineering & Technology, 2009. **32**(12): p. 1992-1999.

507312 An Improved Bubble-Based Drag ModelÂ for Accurate Coarse-Grid Two-Fluid Modeling of Geldart a Powder Bubbling Fluidization

Abstract:

An improved bubble-based energy minimization multi-scale (EMMS) drag model was formulated in which the original bubble-based EMMS drag model at macroscale bed level was extended to microscale sub-grid level for accurate coarse-grid two-fluid modeling (TFM) of Geldart A powder bubbling fluidization. At both the macroscale and microscale levels, the physically heterogeneous structures were modeled by three pseudo phases, i.e., bubble phase, emulsion phase, and inter-phase, where conservation and stability condition equations were applied. Through such scale and phase decompositions, the improved bubble-based EMMS drag model was formulated through a two-step macro-to-micro scheme, where the obtained drag coefficient is related to phase volume fractions, velocities, and accelerations within each coarse grid. Modeling results show that TFM combined with the improved bubble-based EMMS drag model can accurately reproduce the heterogeneous structures for a laboratory-scale Geldart A powder bubbling fluidized bed. At relatively coarse grid, the predicted axial and radial distributions of solid concentration agree very well with experimental measurements. This indicates that our proposed improved bubble-based EMMS drag model incorporated TFM is promising for accurate and economical simulation of engineering-scale Geldart A powder bubbling fluidization.

507860 Numerical Simulation on Fine Particle Transport Behaviour in Electrostatic Precipitators

Abstract:

Electrostatic precipitators (ESPs) are a major air pollution control device to remove particles from treated gas. Extensive research has been done on several aspects such as electrostatics, fluid dynamics, charging mechanism, and particle dynamics. The overall particle-collection efficiency of a modern industrial wire-plate ESP exceeds 99%. However, the removal efficiency of submicron particles by conventional ESPs is as low as 70â€"80% owing to the low charge carried by fine particles and the increase in mobility with decreasing size. Submicron particles possess a high superficial area, thus favouring toxic heavy metals and other pollutants. Therefore, fine particles are collected with much lower efficiency and further research is required to improve this aspect.

In this work, a numerical study on the charging and transport behaviour of fine particles has been carried out based on wire-plate electrostatic precipitators (ESPs) with multiple wire electrodes. The effect of the applied wire voltage, inlet height, and wire distance on particle charging and transport behaviour, and the influence of the precipitator structure on particle trapping are analysed in detail. The 2-D numerical model involves electric field, space-charge density, gas flow, and particle trajectories. The fluid-flow velocity is higher than 1 m/s in this study and the EHD flow has a negligible. A Lagrangian approach for particle transport is utilized with the trajectories of individual particles tracked by solving their equation of motion. The electric field and gas flow are calculated using the finite-volume method. The flow simulation is done using the commercial software FLUENT with other parts coded as user-defined functions. The following main results can be drawn:

(1) It was observed that the Y-displacement of the particles exhibited a universal increase for all particle sizes, especially for larger particles, at low injection positions ($Y < 25$ mm). The larger the particle size, the higher the Y-displacement of particles because larger particle sizes improve both the particle charging level and electric field strength. No significant improvements were found for submicron particles.

(2) Increasing the spacing of the discharge electrode within a certain range and therefore increasing the electric field, it was observed that the particle-trapping effect improved and the outlet level exhibited a universal increase for all particle sizes, especially for large particles.

(3) The ESP models M1 and M2 were used to investigate their influence on the outlet height. Except for particles with a diameter of 2 μm, all other particle sizes showed an increased outlet height in M2 compared to that in M0. In addition, the different discharge-electrode arrangements, especially in M2, exhibited significant effects on the charging and transport behaviour of particles with a diameter of 1 μm.

(4) Particles with an inlet height of 5 and 10 mm were studied in M1 and M2 precipitators. The changed discharge-electrode arrangement improves particle trapping, especially in M2. This is because of the electron trajectories being closer to the discharge electrodes (M2), giving the particles more charge and leading to a greater electric field.

507879 Lattice Boltzmann Simulations of Porous Particulate Flows

Abstract:

The comprehensive understanding of the particle-fluid interaction is very important for fluid-particles system, and thus many studies have been conducted for the particle-fluid system[1-3] of solid impermeable particles. In the real industrial fields such as the catalyst process in chemical engineering[4], the particles usually possess the porous structure which will have effects on the fluid-particles interaction[5-6]. However, there has been little open literature discussing the particle-fluid system of porous permeable particles. Therefore, the objective of the present study is to investigate the effect of the porosity and permeability on the porous particulate flows which are formulated by the volume-averaged governing equations[7] in terms of intrinsic phase average velocity. The lattice Boltzmann equation model (LBE)[8] is used to solve the governing equations. Firstly, we validate the LBE model by the numerical results existing in the literature for the flow around and through one porous particle. Then, the porous particles settling against gravity in a quiescent fluid, i.e., the DKT process[9] for porous particles, is investigated. The effects of the porosity and Darcy number on the porous particulate flows are studied numerically in details. Also, we present the simulation results of the fluidization of 512 porous particles. Our results demonstrate the significance of the porous structure of particles for the particles dynamics behaviors.

Reference

[1] Ladd, A.J.C., 1994. Numerical simulations of particulate suspensions via a discretized Boltzmann equation Part I. Theoretical foundation. J. Fluid Mech., 271, 285-310.

[2] Hill, R.J., Koch, D.L., Ladd, A.J.C., 2001. The first effects of fluid inertia on flows in ordered and random arrays of spheres. J. Fluid Mech. 448, 213-241.

[3] Van der Hoef, M.A., Beetstra, R., Kuipers, J.A.M., 2005. Lattice Boltzmann simulations of low-Reynolds-number flow past mono- and bidisperse arrays of spheres: results for the permeability and drag force. J. Fluid Mech. 528, 233-254.

[4] Zhao, Y.F., Li, H., Ye, M., Liu, Z.M., 2013. 3D numerical simulation of a large scale MTO fluidized bed reactor. Ind. Eng. Chem. Res. 52, 11354-11364.

[5] Noymer, P.D., Glicksman, L.R., Devendran, A., 1998. Drag on a permeable cylinder in steady flow at moderate Reynolds numbers. Chem. Eng. Sci. 53, 2859-2869.

[6] Zhu, Q.Y., Chen, Y.Q., Yu, H.Z., 2014. Numerical simulation of the flow around and through a hygroscopic porous circular cylinder. Comput. Fluids 92, 188-198.

[7] Wang, L., Wang, L.-P., Guo, Z.L., Mi, J.C., 2015. Volume-averaged macroscopic equation for fluid flow in moving porous media. Int. J. Heat Mass Trans. 82, 357-368.

[8] Chen, S.Y., Doolen, G.D., 1998. Lattice Boltzmann method for fluid flows. Annu. Rev. Fluid Mech. 30, 329-364.

[9] Fortes, A.F., Joseph, D.D., Lundgren, T.S., 1987. Nonlinear mechanics of fluidization of beds of spherical particles. J. Fluid Mech. 177, 467-483.

508370 Modeling of a Novel Multi-Particle Collision Model for Gas-Solid FlowsÂ

Abstract:

Fluidized beds are widely applied in the chemical, petrochemical, metallurgical, environmental and energy industries due to their favourable mass and heat transfer characteristics. In past few decades computational fluid dynamics has proven to predict hydrodynamics of fluidized beds. DPM (Euler-Lagrange) [1] and TFM (Euler-Euler) [2] are two most common modelling approaches for gas-solid flow in fluidized beds. In TFM the solid phase itself is treated as a continuum phase interpenetrating with the gas phase. Since the concept of individual particles has disappeared, the effects of particle-particle interactions in the TFM has to be incorporated in a statistically averaged way, via the solid phase pressure and viscosity. TFM is preferred model to study large scale fluidized beds but model suffer from high numerical diffusion. And the KTGF theory in TFM assume elastic collisions and is limited only to spherical particles [2]. In DPM, the gas phase is considered as the continuous phase whereas the particle trajectories are obtained by integrating Newtonâ€™s equations of motion. Particle-particle and particle-wall interactions are explicitly taken into account using various physical models such as the soft sphere model. DPM proves to be accurate modelling of gas-particle flow. However, lagrangian models quickly become computationally too expensive because of the large number of particle interactions. Therefore coarse graining of discrete particles is an emerging model that can handle large number of particles effectively. The new coarse-grained model developments are of vital importance for the advancement of large-scale modelling of gas-particle flows that are flexible to extended for polydispersity, and non-spherical particles. In this study a novel lagrangian multi-particle collision (MPC) method is developed for predicting dense particulate-fluid flow.

The MPC solver is developed in OpenFoam [5], where the fluid phase is treated as a continuum and the solids are modelled by lagrangian particles. The MPC model grew out of the multiphase particle-in-cell (MP-PIC) model [3], that employs a fixed Eulerian grid, and particles are traced in a lagrangian fashion, including a parcel approach where each simulated particle represents a large collection of real particles. Each parcel follows Newtonâ€™s equations of motion, where inter-phase momentum transfer is accounted through mapping and interpolating quantities back and forth between parcel locations and the eulerian grid. Particle collisions in MP-PIC are not fully resolved but are derived from particles stresses that are theoretically calculated from Harris and Crighton [4] stress closure. In the MPC model, we implemented a functional form for the particle stress tensor where the coefficients are obtained computationally by means of DPM simulations. The idea is to extend the model from spherical particles to non-spherical particles, for which theoretical particles closures are currently not available in the literature. The viscous stress and random forces are typically ignored in MP-PIC model due to their insignificant effect [3]. Importantly, we incorporated viscous stress and random forces representing scattering of particles by collisions, which are obtained from DPM simulations by measuring the actual variance. Details of model development and validation is presented in this study.

Results from MPC model are compared with MPPICFoam and, with DPM simulations from DEM-LIGGGHT [6], and MFiX [7] code respectively. Pseuduo-2D fluidized beds of dimension 0.01x0.15x0.45 m is considered for comparison with experiments [8]. Comparison is made on the basic of pressure drop, void fraction, bubble dynamics and solids motion. We observed that MPC is in good agreement with DPM, and experiments, when compared to MPPICFoam. Unrealistic over-packing of solids is predicted in MPPIC due to limitation of Harris and Crighton [4] model and the velocity field updates, whereas in MPC these limitations overcomes and no over packing of particles is predicted. Solid circulations are in good agreement with DPM results, where solid moving above in the center and downward near to the wall. Time-average porosity plots revels the path of bubble motion and is in close agreement with results of DPM. In conclusion a novel multi-particle collision model is developed and validated to study gas-solid fluidized beds. The model is efficient to handle large number of particles. In future study, we plan to extend the model for dense flow of non-spherical particles.

Acknowledgments

The authors thank the European Research Council for its financial support under its consolidator grant scheme, contract no. 615906 (NonSphereFlow).

References

1. N.A. Patankar, D.D. Joseph, 2001. Modeling and numerical simulation of particulate flows by the Eulerian Lagrangian approach, International Journal of Multiphase Flow, Volume 27, Issue 10, October 2001, Pages 1659-1684.
2. Gidaspow, D., 1994. Multiphase flow and fluidization continuum and kinetic theory description, Academic Press, Boston.
3. Andrews, M.J., O'Rourke, P.J., 1996. The multiphase particle-in-cell (MP-PIC) method for dense particulate ows, Int. J. Multiphase Flow 22, 379-402.
4. Harris, S.E., Crighton, D.G., 1988. Solutions, solitary waves and voidage disturbances in gas-fluidized beds, J. Fluid Mech. 195, 437.
5. OpenFOAM, 2015. The Open Source CFD Toolbox, Ver. 3.0.0. ESI Group, Paris, France.
6. https://www.cfdem.com/liggghtsr-open-source-discrete-element-method-particle-simulation-code
7. https://mfix.netl.doe.gov/mfix/
8. M. Goldschmidt, 2001. PhD thesis, Hydrodynamic Modelling of Fluidised Bed Spray Granulation.

508490 Dpm Analysis of Large Fluidized Catalytic Cracking (FCC) Reactors

Abstract:

DPM Analysis of Large Fluidized Catalytic Cracking (FCC) Reactors

Abstract:

Computational Fluid Dynamic (CFD) modeling is a powerful tool for equipment design and development and is widely used at UOP, particularly for the reactor and regenerator design of FCC process unit. Through CFD modeling, a technical solution was developed for a large capacity FCC process unit purposing an elevated amount of propylene production.

A key challenge in FCC process design is how to optimize the size of the reactor vessel, which is a function of the number and size of reactor cyclones (catalyst separationdisengagers). UOP developed a new cyclone cluster configuration via CFD study of the reactor internals, including UOP's VSS™ riser termination device and reactor cyclones, which reduce the reactor size significantly compare to the conventional design, while maintaining effective catalyst / vapor separation and mechanical reliability. The objective of the current CFD study is to evaluate the catalyst separation efficiency and performance of the new riser termination device technology, which accommodate more cyclones in a smaller reactor vessel.

To better understand the flow pattern, flow distribution and the gas residence time inside the UOP VSS™ , the analysis was broken down into two sections. Part 1 focused on single gas phase only and in part 2 the particles behavior is studied using the Discrete Phase Model (DPM) in Fluent.

The single gas flow analysis results showed that the flow in the VSS chamber section has the same rotational pattern as in conventional FCC/VSS units, where the catalyst will separate from the gas products. Also the gas flow distribution at each chamber arm is uniform, which is critical to catalyst separation efficiency.

The DPM analysis confirmed an equal catalyst particle distribution through all the cyclones and a high efficiency separation in the riser termination device. Having this information from CFD allows for the new design to move forward.

509623 Catalytic Reactor Design Using Multiphase CFD

Abstract:

Catalytic Reactor Design Usng Multiphase Cfddimitri Gidaspow the 2011 Book , "Computational Gas-Solids Flows and Reacting Systems ", Edited By S. Pannala, M. Syamlal and Tom O'Brien ,Engineering Science Reference, Shows the Progress We Have Made in Using Multiphase CFD in Modeling Fluidized Bed Chemical Reactors and Other Multiphase Systems, Such As the Pyroplastic Flow Emitted from Volcanoes. in This Presentation the Design of Two Types of Catalytic Reactors Will be Described: A Fischer_ Tropsch Slurry Bubble Column Reactora Polymerization Reactor the CFD Design of a New Slurry Bubble Column Reactor Was Already Described By Gidaspow ,Et Al in Chemical Engineering Science 134,784-799 (2015).in This Presentation the Design Will be Extended to Obtain 100, 000 Barrels of Diesel per Day, Twice That of the Largest Commercial Reactors. the Production of Synthesis Gas Using Oxygen or Carbon Dioxide in Catalytic Fluidized Beds Needed for the Fischer-Tropsch Reactor Will Also be Discussed Briefly. for the Polymerization Reactor Our CFD Code Described in " Computational Techniques " By D. Gidaspow and V. Jiradilok , Nova Science, 2009 Was Modified for Polymerization Reactors. Such Reactors and Their CFD Modeling Are Described in a Chapter in Pannala's Et Al Book from Professor Rodney Fox's Group. They Point out That the Fine Particle (80 microns) Catalyst Radial Distribution, Is a Remaining Issue the Computations to be Presented Will Show That the Catalyst Will Have a High Concentration at the Reactor Walls for Various Operating Conditions. This Leads to a High Rate of Reaction at the Reactor Walls and May Result in Melting of the Polymers and Sheet Formation , Forcing Reactor Shut-Down. in Some of Our Simulations There Is Large Bubble and Slug Formation Similar to Those Mentioned in Fox's Chapter. Acknowledgement : I Thank Dr. Mayank Kashyap of SABIC for a Valuable Discussion and His Papers " Application of Multiphase Flow CFD in the Gas Phase Polymerization Process " Fluidization XV. May 24,2016 and a Related Paper Presented at Doe NETL, July 2016

509627 Simulation of Large Particle Turbulent Fluidization in Riser Reactors By Coarse Grain DEM-CFD

Abstract:

The complex two-phase flow in the riser is an important element of many industrial circulating fluidized bed reactors. In some cases, e.g. gas-solid polymerization, relatively large particles are transported upward in a turbulent regime, in which the high gas velocity results mostly from heat removal requirements. Strong solids acceleration, radial and axial velocity profiles for both the gas and solid, intense clustering and inhomogeneities make it impossible to reliably predict the riser head or local bed density profile. Detailed numerical simulations proved capable of showing some of the peculiarities of the flow, but limitations in the scale, lack of grid independence, need for special drag corrections and sub-grid models still hinder confident use of such tools in design and scale-up, particularly in industrial applications characterized by particle size and Reynolds numbers that are not well covered by the available scientific literature.

In the present contribution, coupled Discrete Element Method-Computational Fluid Dynamics (DEM-CFD) simulations are carried out using a coarse graining technique for the solid particles. According to this method, small number of particles are grouped into a computational particle (*parcel*) and represented by it as a spherical "lump," to reduce the computational cost of the simulations and allow lab-scale systems to be represented in full detail. The investigation focuses on the influence of the coars-graining degree, i.e. the number of real particles represented by a *parcel*, on the simulated flow hydrodynamics both in terms of its advantages (a quantification of the computational savings) and disadvantage, primarily the real capability of the method to resolve the behaviour observed and produce results in line with experimental findings. The gas and solid motion are studied, with the aim to characterize the clustering phenomena observed during the up-flow and its effect on the pressure drop. Assuming prescribed gas and solid inlet velocity fields, the potential tendency to show core-annulus flow, the intensity of solids clustering and the local bed density within the riser are examined at different number of particles per *parcel*.

509640 Numerical Study of Particles Shape Effects on Solid-Liquid Fluidizations

Abstract:

Numerical study of particles shape effects on solid-liquid fluidizations

Esmaeil Abbaszadeh Molaei, Zongyan Zhou*

Laboratory for Simulation and Modelling of Particulate Systems, Department of Chemical Engineering, Monash University, Clayton, VIC 3800, Australia

Solid-liquid fluidizations have been observed in many industries due to their high heat and mass transfer rate, high operation flexibilities, and reduced back mixing of fluid and solid phases. In most of these multiphase operations, particle properties such as size, density, and shape may experience significant changes during the process because of attrition, coalescence, comminution or chemical reactions. The difference in particle properties can result in significant segregation problem in liquid fluidized beds. To gain an understanding of solid-liquid fluidisation behaviour, most existing studies, either experimentally or numerically, have focused on solid flow behaviour with mono-sized particles or binary and ternary mixtures with different particle sizes or densities. In spite of substantial studies on the hydrodynamics of liquid fluidized beds, the effects of particles shape on the solid-liquid flow behaviour are still less reported. Except for two physical experiments, the previous studies considered particles as spheres, and few studies have been remarked on liquid fluidization of non-spherical particles. However, particles are generally irregular in most of the processes such as ore beneficiations using liquid-solid fluidized beds separator. Particle shape can cause different flow behaviour in liquid fluidizations rather than that observed for ideal spherical particles. Nonetheless, the answers to some fundamental questions such as how and why particle shape affects flow phenomena are not clear, and hence still poorly understood.

Therefore, in this work, a combination of computational fluid dynamics (CFD) and discrete element model (DEM) [CFD-DEM] is used to examine the effects of particle shape on the flow characteristics in liquid fluidizations. In the simulations, ellipsoidal particles are used as they can represent a wide range of particle shapes from oblate to prolate particles. Different aspect ratios are employed, and results are analysed mainly in terms of particle flow patterns, pressure drop, and bed expansion. In addition, force and particle orientation analysis is also conducted to examine the effect of particle shape and explain the occurrence of entrainment phenomenon.

The results show that ellipsoidal particles intend to enter to freeboard region and entrainment phenomenon occurs, and consequently, the interface between the bed and freeboard cannot be observed clearly. In addition, the pressure gradient variation with bed height is uniform for spheres; however, this uniformity is replaced with â€œdistorted parentheses-shapeâ€ profile for disc-shape and elongated particles. Finally, bed expansion analyses show that spherical particles have the lowest bed expansion. On the other hand, disc-shape particles intend to expand more significantly than spherical and elongated particles. This causes the entrainment phenomenon which has been found to be due to the effects of particles shape on particles terminal velocity.

* Corresponding author. Tel: 61 3 9905 0846; email: zongyan.zhou@monash.edu.

514318 CFD Investigation on Gas-Solids Flow and Heat Transfer in Two Fluidized Catalyst Cooler

Abstract:

Our previously experimental work showed that the internal circulation of solids in an annular catalyst cooler (ACC) had an ability to enhance the heat transfer between bed and wall, comparing with the base catalyst cooler (BCC). The current work simulated in detailed hydrodynamics of gas-solids flow and heat transfer to advance our understanding of the ACC by using the two-fluid model. The effective particle diameter $d_p^{\text{âŽ}}$ was determined to 170 Î¼m and the specularity coefficient Ï† was 0.3 in the 3D simulations. The catalyst cooler can be divided to four distinct regions including draft tube region, annulus region, bottom region and particle diffluence region according to the distributions of solids volume fraction and solids velocity. The effect of gas-solids flow on bed-to-wall heat transfer was compared in the different regions of the BCC and ACC. And the effects of operational gas velocity and bed structure were discussed. The results showed that the proper selection of the ratio of gas velocities and the position of the heat transfer tube were needed to form a stable internal solids circulation in the ACC. The ACC had a combined hydrodynamic feature of up-flow and down-flow catalyst coolers with bigger solids volume fraction and smaller particle resident time, which are beneficial for improving the heat transfer between solids and wall.

Hydrodynamics and mixing characteristics of a new-type particle mixer

Mengxi Liu, Chunxi Lu, Zhenliang Meng

State Key Laboratory of Heavy Oil Processing, China University of Petroleum, Beijing 102249, PR China

E-mail: mengxiliu@sina.com (M. Liu)

Abstract:

In order to further enhance the particle mixing in the internally circulating fluidized bed, especially along the radial direction, a new-type particle mixer with slots on the draft tube was proposed in the present work. The hydrodynamics of the mixer was investigated by using a multi-scale computational fluid dynamics with the EMMS drag. The prediction in terms of cross-sectionally averaged solid holdup and solid velocity were compared to experimental data for validation. The results show that in draft tube a low pressure region can be observed near the slots and in annulus the pressure in the vicinity of the slots seems larger than that in the region between the adjacent slots, which induces horizontal particle immigration and mixing. Nearly 25.4% to 29.7% of gas in the draft tube bypasses and flows through the slots into the annulus. The ratio of the mass flux of the tracer gas bypassing into the annulus to that of total tracer gas is approximately equal to 0. With increase of the superficial gas velocity in the draft tube from 0.3 m/s to 0.45 m/s, the total solid mass flux increases from 4.87 kg/s to 10.86 kg/s, while the mass flux through slots decreases from 2.69 kg/s to 0.52 kg/s.

1. Introduction

The characteristics of particle flow and mixing have a considerable influence on the mass and heat transfer, as well as the product quality. According to the previous study, the mixing degree of particles along the axial and radial direction is quite different [1-5]. Mostoufi and Chaouki [4] discovered that the diffusivities varied from 3.3×10^{-3} m^2/s to 5.6×10^{-2} m^2/s along the axial direction and 2.6×10^{-4} m^2/s to 1.5×10^{-3} m^2/s along the radial direction. Berruti et al. [6] also revealed that the particle mixing rate for radial direction was an order of magnitude smaller than for the axial direction. Therefore, the weak particle flow along the radial direction greatly limits the mixing ability of a fluidized bed. Internally circulating fluidized bed (ICFB) has been widely recognized as an efficient fluidized bed reactor, in which orderly and macroscopic particle circulation occurs. Compared with the conventional fluidized bed, ICFB offers multiple particle circulation between the draft tube and the annulus, which leads to relatively strong particle convection, diffusion and shear, and thereby improves the particle mixing performance along the radial direction [7]. In order to further enhance the particle mixing for some processes in which a uniform particle mixing is required, a new-type particle mixer with slots on the draft tube was proposed in the present work.

With increase of the computational power, the computational fluid dynamics (CFD) provides an additional tool to predict particle flow and mixing in the local region, which is difficult with the experimental methods. In particular, the gas-solid flow in a gas-solid fluidized bed can be accurately described by the energy minimization multi-scale (EMMS) model [8]. In this paper, the hydrodynamics of the mixer was investigated by using a multi-scale CFD based on the EMMS drag model [9, 10].

2. Experimental apparatus, Materials and Operating conditions

The structure of the particle mixer is shown in Fig.1. The mixer has a height of 5 m and an inner diameter of 286 mm. A draft tube with a height of 1 m and an inner diameter of 215 mm is coaxially mounted into the column. The four rectangular slots of 80×40 mm are symmetrically opened on the draft tube at a height of *h*=860 mm, in order to strengthen the particle radial flow. A gas distributor plate and a ring distributor are installed below the draft tube and the annulus respectively. The height of the gap between the gas distributor plate and the draft tube is 60 mm. For the sake of the convenience, the mixer is divided into six zones: the pre-mixing zone, feed influence zone, draft tube zone, gas-solid separation zone, annulus zone and the bottom zone. A perforated plate is mounted at the bottom of the pre-mixing zone, which is connected with a central downcomer, extending into the draft tube. Particles are first mixed in the pre-mixing zone and then the pre-mixed particles flow downward through the central downcomer and into the feed influence zone. In the feed influence zone, the pre-mixed particles are mixed with the particles moving upward in the draft tube, flow upward and circulate between the draft tube and the annulus. In the annulus, particles move downward to the bottom. Then a part of particles flow out of the mixer and the other part flow horizontally into the draft tube and recirculate.

The solid particles used in the experiment were FCC catalyst (ρ_p=1498 kg/m^3), with an average diameter of 76 μm. Air (ρ_g=1.225 kg/m^3) was introduced into the mixer to fluidize the particles. The superficial gas velocity ranged from 0.3 to 0.45 m/s in the draft tube (based on the cross-sectional area of the draft tube) and remained at 0.05m/s in the annulus (based on the cross-sectional area of the annulus). The mass flux of cold particles remained at 13.8 kg/m^{2}s. As we only investigated the influence of slots on the flow and mixing of particles in this paper, no hot particles were induced into the mixer.

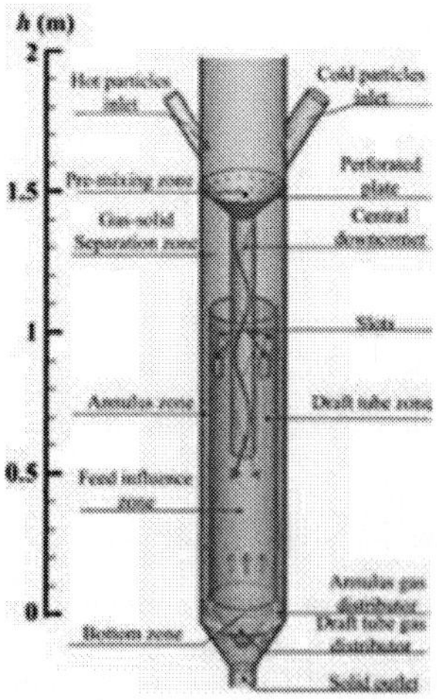

Fig. 1 Schematic diagram of the new-type particle mixer

3. CFD model and simulation method

The hydrodynamics of the mixer was performed using the Eulerian- Eulerian of Ansys Fluent

®13.0. The governing equations [11] include a set of mass and momentum equations for each phase, as shown in Table 1.

A 3-D mesh containing 580,000 cells was generated by using Gambit®2.4. The no-slip boundary was assigned for the gas phase and the partial-slip boundary condition developed by Johnson and Jackson [12] was used for the solid phase. In order to ensure a steady-state operation of the mixer, simulations were first performed for 30 s, and then the time-averaged data was analyzed for another period of 10 s. Other parameters settings in the model are shown in Table 2.

Table 1 Governing equations

Gas-phase Continuity equation $\frac{\partial}{\partial t}(\varepsilon_g\rho_g)+\nabla\cdot(\varepsilon_g\rho_g\boldsymbol{u}_g)=0$

Solid-phase Continuity equation $\frac{\partial}{\partial t}(\varepsilon_s\rho_s)+\nabla\cdot(\varepsilon_s\rho_s\boldsymbol{u}_s)=0$

Gas-phase momentum equation $\frac{\partial}{\partial t}(\varepsilon_g\rho_g\boldsymbol{u}_g)+\nabla\cdot(\varepsilon_g\rho_g\boldsymbol{u}_g\boldsymbol{u}_g)=-\varepsilon_g\nabla p+\nabla\cdot\boldsymbol{\tau}_g+\varepsilon_g\rho_g\boldsymbol{g}-\beta(\boldsymbol{u}_g-\boldsymbol{u}_s)$

Solid-phase momentum equation $\frac{\partial}{\partial t}(\varepsilon_s\rho_s\boldsymbol{u}_s)+\nabla\cdot(\varepsilon_s\rho_s\boldsymbol{u}_s\boldsymbol{u}_s)=-\varepsilon_s\nabla p-\nabla p_s+\nabla\cdot\boldsymbol{\tau}_s+\varepsilon_s\rho_s\boldsymbol{g}+\beta(\boldsymbol{u}_g-\boldsymbol{u}_s)$

Granular temperature equation

$$\frac{3}{2}[\frac{\partial}{\partial t}(\varepsilon_s\rho_s\Theta_s)+\nabla\cdot(\varepsilon_s\rho_s\Theta_s\boldsymbol{u}_s)]=(-p_s\boldsymbol{I}+\boldsymbol{\tau}_s):\nabla\boldsymbol{u}_s+\nabla\cdot(\kappa_s\nabla\Theta_s)-\gamma_s-3\beta\Theta_s$$

Gas-phase stress $\boldsymbol{\tau}_g=\varepsilon_g\mu_g[\nabla\boldsymbol{u}_g+(\nabla\boldsymbol{u}_g)^T]-\frac{2}{3}\varepsilon_g\mu_g(\nabla\cdot\boldsymbol{u}_g)\boldsymbol{I}$

Solid-phase stress $\boldsymbol{\tau}_s=\varepsilon_s\mu_s[\nabla\boldsymbol{u}_s+(\nabla\boldsymbol{u}_s)^T]+\varepsilon_s(\lambda_s-\frac{2}{3}\mu_s)(\nabla\cdot\boldsymbol{u}_k)\boldsymbol{I}$

Solid-phase pressure $p_s=\varepsilon_s\rho_s\Theta_s[1+2(1+e_s)\varepsilon_s g_0]$

Solid-phase shear viscosity $\mu_s=\mu_{s,col}+\mu_{s,kin}+\mu_{s,fr}$

Where $\mu_{s,col}=\frac{4}{5}\varepsilon_s\rho_s d_p g_0(1+e_s)\sqrt{\Theta_s/\pi}$, $\mu_{s,kin}=\frac{10\rho_s d_p\sqrt{\Theta_s/\pi}}{96\varepsilon_s(1+e_s)g_0}[1+\frac{4}{5}g_0\varepsilon_s(1+e_s)]^2$, $\mu_{s,fr}=\frac{p_s\sin\phi}{2\sqrt{I_{2D}}}$

Radial distribution function $g_0=[1-(\varepsilon_s/\varepsilon_{s,max})^{1/3}]^{-1}$

Diffusion coefficient $\kappa_s=\frac{150\rho_s d_p\sqrt{\Theta_s/\pi}}{384(1+e_s)g_0}[1+\frac{6}{5}g_0\varepsilon_s(1+e_s)]^2+2\rho_s\varepsilon_s^2 d_p(1+e_s)g_0\sqrt{\Theta_s/\pi}$

Collisional energy dissipation $\gamma_s=\frac{12(1-e_s^2)\varepsilon_s^2\rho_s g_0\Theta_s^{3/2}}{d_p\sqrt{\pi}}$

Heterogeneity index $H_d=\frac{\beta_e}{\beta}$, $\beta_e=\beta H_d$

Drag based homogeneous structure

$$\beta=\begin{cases}\frac{3}{4}C_d\frac{\varepsilon_s\varepsilon_g\rho_g|\boldsymbol{u}_g-\boldsymbol{u}_s|}{d_p}\varepsilon_g^{-2.65}, & \varepsilon_g>0.8\\ 150\frac{\varepsilon_s^2}{\varepsilon_g}\frac{\mu_g}{d_p^2}+1.75\varepsilon_s\frac{\rho_p}{d_p}|\boldsymbol{u}_g-\boldsymbol{u}_s|, & \varepsilon_g\le 0.8\end{cases}$$

Where $C_d=\begin{cases}0.44 & \mathrm{Re}_s\ge 1000\\ \frac{24}{\mathrm{Re}_s}(1+0.15\mathrm{Re}_s^{0.687}), & \mathrm{Re}_s<1000\end{cases}$, $\mathrm{Re}_s=\frac{\varepsilon_g\rho_g d_p|\boldsymbol{u}_g-\boldsymbol{u}_s|}{\mu_g}$

Table 2 Simulation Parameters

Gas-solid model	Eulerian-Eulerian
Solids Pressure	Lun et al.
Particle-wall specularity coefficient	0.5
Gas-wall friction coefficient	No-slip
Pressure-Velocity Coupling	Simple
Gradient discretization	Least squares cell based
Momentum discretization	Second-order upwind
Volume fraction discretization	Quick
Max. number of iterations per time step	40
Pressure outlet (atm)	1
Inventory (kg)	56
Incipient solid holdup	0.6

4. Results and discussion

4.1 Model validation

Fig. 2 gives the distribution of cross-sectionally averaged solid holdup and particle velocity in the mixer. As shown in Fig. 2(a), the solid holdup is larger in the bottom zone (h=0.04 m) and due to the accelerated upward motion of particles, the solid holdup decreases with increase of the axial height (when $h \leq 0.16$ m). The cross-sectionally averaged solid holdup is nearly constant in the draft tube and gas-solid separation zone of $0.4 \leq h \leq 1.25$ m. The cross-sectionally averaged particle velocity is shown in Fig. 2(b). The particle velocity is smaller in the bottom zone and keeps almost invariable in the draft tube and the gas-solid separation zone. Compared with the experimental data, the particle velocity seems slightly higher, while the solid holdup in the simulation agrees reasonably with the experimental results.

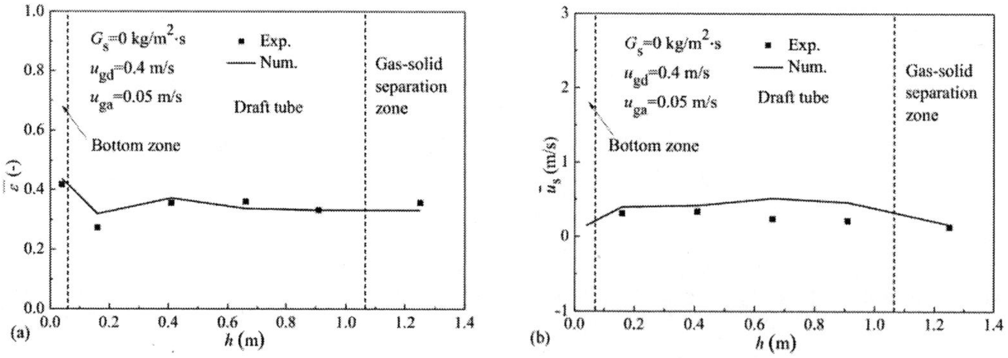

Fig. 2 Distribution of cross-sectionally averaged solid holdup and particle velocity as a function of axial heights

4.2 Pressure distribution near the slots

Fig. 3 shows the time-averaged contours of static pressure on the cross-section of h=0.9 m, the cross-section on the middle of the slots. It is seen that in the draft tube a low pressure region can be observed near the slots, smaller than the pressure in the region between the adjacent slots. This pressure drop forced particles inside the draft tube to move towards the slots and pass through them, which significantly facilitates the mixing of the particles along the radial direction. In the annulus the pressure in the vicinity of the slots seems larger than that in the region between the adjacent slots, inducing horizontal particle immigration and mixing.

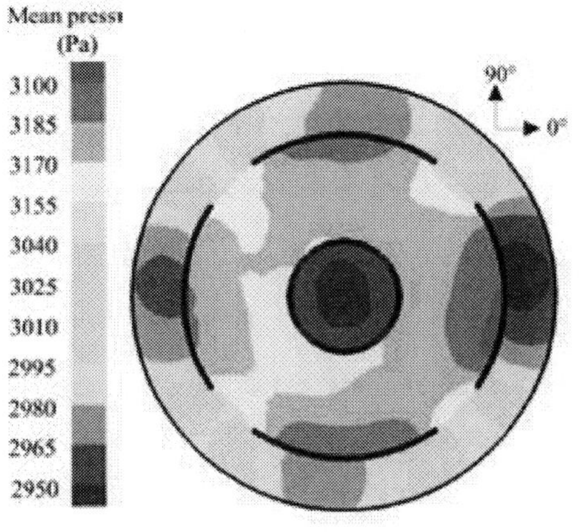

Fig. 3 Time-averaged contours of static pressure at the cross-section of the slots (u_{gd} =0.4 m/s, G_s=13.8 kg/m^{2}s, h=0.9 m)

4.3 Gas bypassing

In the present work, the gas bypassing behavior was investigated by CFD simulation, and a tracer gas having identical physical properties with air was defined. After the mixer running stably, the tracer gas was introduced into the mixer from the gas distributor plate to replace the original gas. A cross section (h=0.06 m) on the exit of the annulus was selected and the mass flux of the tracer gas through the cross section was registered. The results are shown in Fig. 4. f_B refers to the ratio of the mass flux of the tracer gas through the cross section (h=0.06 m) to that of total tracer gas. It is seen that f_B varies between -0.05 and 0.05 with a frequency of 10 Hz, indicating an unstable gas bypassing from the draft tube to annulus. This suggests that the gas bypassing into the annulus was finally brought out of the annulus by the circulating particles. The average f_B is approximately equal to 0, much smaller than that in a conventional ICFB (usually within 15%) [13], because massive particle circulation in the mixer makes it difficult to bypass from the draft tube to annulus.

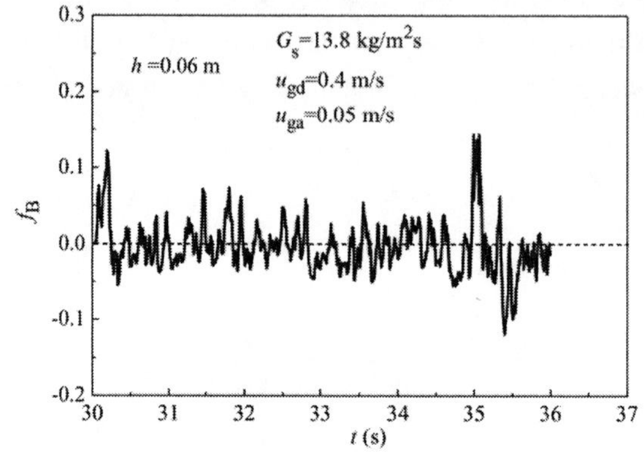

Fig. 4 Gas bypassing from draft tube to annulus in the bottom zone

As discussed previously, in the mixer there exists a pressure gradient from the draft tube to the vicinity of the slots, and then the annulus. Such a pressure difference leads to horizontal flows of particles and gas from the draft tube to the slots and then to the annulus. In the mixer the upside and downside of the slots is located at h=0.94 m and 0.86 m, respectively. Mass flux of the gas bypassing through the slots was obtained by monitoring the gas flux through the lower cross section (h=0.7 m) and the upper cross section (h=1 m) in the draft tube. Fig. 5 shows the gas mass flux under different operating conditions. The total gas mass flux in the draft tube increases with increasing gas velocity in the draft tube. With increase of the gas velocity from 0.3 to 0.45 m/s, the mass flux of the gas bypassing through the slots increases from 0.157 to 0.221 kg/m^{2}s. But the mass flux of bypassing gas does not increase linearly with the gas velocity. It is also seen that the mass ratio of the gas through the slots to that in the draft tube drops from 29.7% to 25.4%.

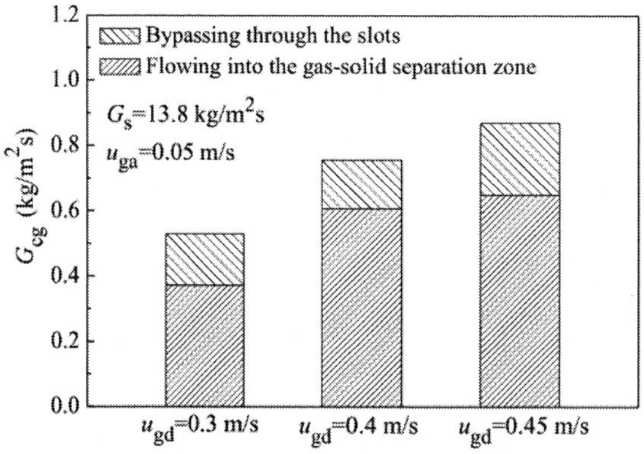

Fig. 5 Gas mass flux through the slots and the gas-solid separation zone as a function of superficial gas velocity in the draft tube

4.4 Particle mass flux

Fig. 6 shows the particle mass flux in the mixer. It is seen that the mass flux increases with increase of the superficial gas velocity in the draft tube. The mass flux through the slots accounts for about 50% of the total mass flux when u_{gd}=0.3 m/s. Because the axial particle velocity rose with increase of the superficial gas velocity from 0.3 m/s to 0.45 m/s in the draft tube, making it difficult to change the direction of the particles. Meanwhile, the radial particle velocity above the draft tube increases. Consequently, with the increase of the superficial gas velocity in the draft tube , the mass flux through the slots considerably decreased，from about 55.3% of the total mass flux (about 2.69 kg/s) at u_{gd}=0.3 m/s to only 4.8% (about 0.52 kg/s) at u_{gd}=0.45 m/s. Meanwhile, the total particle mass flux in the mixer increased from 4.87 to 10.86 kg/s as u_{gd} increased from 0.3 to 0.45 m/s.

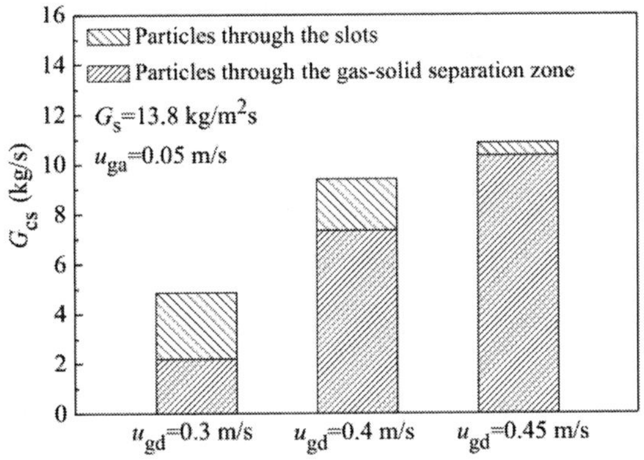

Fig. 6 Particle mass flux through the slots and the gas-solid separation zone as a function of superficial gas velocity in the draft tube

5. Conclusions

The characteristics of particle flow and mixing have a considerable influence on the mass and heat transfer, as well as the product quality. In order to further enhance the particle mixing in the ICFB, especially along the radial direction, a new-type particle mixer with slots on the draft tube was proposed in the present work. The hydrodynamics of the mixer was investigated by using a multi-scale computational fluid dynamics with the EMMS drag. The prediction in terms of cross-sectionally averaged solid holdup and solid velocity were compared to experimental data for validation. The results show that in draft tube a low pressure region can be observed near the slots and in annulus the pressure in the vicinity of the slots seems larger than that in the region between the adjacent slots, which induces horizontal particle immigration and mixing. Nearly 25.4% to 29.7% of gas in the draft tube bypasses and flows through the slots into the annulus. The ratio of the mass flux of the tracer gas bypassing into the annulus to that of total tracer gas is approximately equal to 0. With increasing the superficial gas velocity in the draft tube from 0.3 m/s to 0.45 m/s, the total solid mass flux increases from 4.87 kg/s to 10.86 kg/s, while the mass flux through slots decreases from 2.69 kg/s to 0.52 kg/s.

Notation

G_s	solid mass flux (kg/m^2 s)	G_{cg}	gas mass flux in the mixer (kg/m^2 s)
G_{cs}	solid mass flux in the mixer (kg/s)	h	height (m)

p	pressure (Pa)	R	radius of the reactor (mm)
r	radial position of measurement points (mm)	t	time (s)
u_{ga}	superficial gas velocity in the annulus (m/s)	u_{gd}	superficial gas velocity in the draft tube (m/s)
$\bar{u}$	cross-sectionally averaged axial particle velocity (m/s)		

Greek

ε_g	voidage	ε_s	solid holdup
$\bar{\varepsilon}_s$	cross-sectionally averaged solid holdup	ρ	density (kg/m^3)

References

[1] F. Niklasson, H. Thunman, F. Johnsson, B. Leckner, Estimation of solids mixing in a fluidized-bed combustor, Ind Eng Chem Res, 41 (2002) 4663-4673.

[2] L.H. Shen, M.Y. Zhang, Effect of particle size on solids mixing in bubbling fluidized beds, Powder Technol, 97 (1998) 170-177.

[3] L.H. Shen, M.Y. Zhang, Y.Q. Xu, Solids Mixing in Fluidized-Beds, Powder Technol, 84 (1995) 207-212.

[4] N. Mostoufi, J. Chaouki, Local solid mixing in gas-solid fluidized beds, Powder Technol, 114 (2001) 23-31.

[5] D.Y. Liu, X.P. Chen, Lateral solids dispersion coefficient in large-scale fluidized beds, Combust Flame, 157 (2010) 2116-2124.

[6] F. Berruti, D.S. Scott, E. Rhodes, Measuring and Modeling Lateral Solid Mixing in a 3-Dimensional Batch Gas Solid Fluidized-Bed Reactor, Can J Chem Eng, 64 (1986) 48-56.

[7] M.M. Fang, K. Luo, S.L. Yang, K. Zhang, J.R. Fan, Computational Fluid Dynamics-Discrete Element Method Investigation of Solid Mixing Characteristics in an Internally Circulating Fluidized Bed, Ind Eng Chem Res, 52 (2013) 7556-7568.

[8] J. Li, M. Kwauk, Particle-fluid Two-phase Flow: the Energy-minimization Multi-scale Method, Metallurgical Industry Press, Beijing, 1994.

[9] Z.S. Shi, W. Wang, J.H. Li, A bubble-based EMMS model for gas-solid bubbling fluidization, Chem Eng Sci, 66 (2011) 5541-5555.

[10] K. Hong, Z.S. Shi, W. Wang, J.H. Li, A structure-dependent multi-fluid model (SFM) for heterogeneous gas-solid flow, Chem Eng Sci, 99 (2013) 191-202.

[11] D. Gidaspow, Multiphase flow and fluidization: Continuum and Kinetic Theory Descriptions, Academic Press, Boston, 1994.

[12] P.C. Johnson, R. Jackson, Frictional–collisional constitutive relations for granular materials, with application to plane shearing, J. Fluid Mech. 176 (1987) 67–93.

[13] B.H. Song, Y.T. Kim, S.D. Kim, Circulation of solids and gas bypassing in an internally circulating fluidized bed with a draft tube, Chem Eng J, 68 (1997) 115-122.

514498 Power Spectral Density Analysis of Pressure Signal in 18 m Circulation Fluidized Bed Riser

Abstract:

Studies on the gas-solids flow characteristics using Power Spectral Density based on the pressure data from different axial locations in a riser of diameter 80 mm and height 18 m. Results show that the pressure signal fluctuation is mainly affected by gas-solids interaction, which included gas-solids friction, particle-particle collision and particle-particle aggregation. With the increasing solids circulating rate, frequency of particle interactions including particle-particle collision and particle-particle arrogation become strong and the pressure fluctuation also increased. It can be seen that the pressure signal fluctuation intensity decreased gradually with the increasing axial position of the riser. The low frequency but high power phenomenon nearly disappears in the fully developed zone near the top of the riser. Based on the results, it can be concluded that the gas-solids interaction is stronger at the bottom of riser. On the other hand, the gas-solid interaction is more homogeneous, and the pressure fluctuation is relatively weak in the fully developed zone.

514514 Characteristics of Pressure Fluctuations in Particle-Transport Inclined Pipe of a Circulating Fluidized Bed

Abstract:

In a large circulating fluidized bed device, the characteristics of pressure fluctuation in a particle-transport inclined pipe under different solid circulating flux were investigated based on measuring the dynamic pressure of gas-solid two-phase flow by using FCC equilibrium catalyst as the particles. The experimental results showed that with increase of particle mass flux, the flow patterns of FCC particles above and below the butterfly valve in the inclined pipe both successively changed, and the fluctuation amplitude of dynamic pressure increased. The standard deviation of the dynamic pressure below the butterfly valve possessed a linear relation with the particle mass flux, which could be used to characterize the real-time dynamic pressure in the inclined pipe. By wavelet analysis, it is indicated that there exists a main frequency caused by inclined pipe flow structures in the pressure fluctuations above and below the butterfly valve, besides, there is a secondary frequency in the pressure fluctuation below the butterfly valve caused by gas turbulence and gas-solid interactions. With the increase of particle mass flux, the wavelet energy ratio of the secondary frequency gradually decreased. The fluctuation pressure caused by discharge in the inclined pipe is a low frequency pressure and the excitation source of inclined pipe vibration.

508323 Dynamic Modeling of Attrition and Reactions in Circulating Fluidized Bed Reactors

Abstract:

Fluidized bed reactors are widely applied in chemical processes due to the excellent gas-solid contact. A problem that occurs in fluidized beds is the mechanical stress that particles are exposed to during the fluidization, which causes particle attrition. Additionally, fluidized beds are often operated in batch mode, which will cause ever-changing operating conditions with time. This makes fluidized bed modeling especially challenging, because the distributed particle properties, like size, wetness or reaction state, have to be tracked over the whole course of the simulation.

At the Institute of Solids Process Engineering at Hamburg University of Technology, novel process simulation software named DYSSOL is developed, particularly tailored for solids processes. The software is based on the flowsheeting concept. This means that chemical unit operations, like for example chemical reactors or separators, are put into process modules, which are interconnected into a process flowsheet and then mathematically solved. To make dynamic simulations available, new dynamic models had to be proposed for the involved equipment. The models themselves are based on empirical and semi-empirical correlations for hydrodynamics and kinetic particle reaction models are integrated.

The dynamic flowsheeting software was then used to simulate a system of interconnected fluidized bed reactors, which are used for Chemical Looping Combustion of solid fuels. This process is especially interesting, because heterogeneous reactions take place. Further, the process is a reactor-regenerator system causing the reactive solid to be oxidized and reduced and, hence changing its conversion state. Attrition plays a major role for the cost effective operation of the process.

In a small lab scale bubbling bed reactor, the jet and bubbling attrition rate constants for the solid oxygen carrier were experimentally determined under ambient conditions. In a small stand-alone cyclone the cyclone attrition was investigated. Furthermore, the reaction kinetics of the solid material with fuel gases was experimentally determined. All this information was included into the flowsheet simulation of a pilot scale circulating fluidized bed reactor, which is actually operated in our institute to deliver validation data.

The dynamic flowsheet simulation of the pilot plant, which was enhanced with kinetic and attrition information of smaller scale facilities, was conducted over 1000 seconds. A fuel load change was simulated and the chemical conversion was tracked. Over the whole simulation, the attrition and small particles leaving the reactor system was also covered. When comparing the simulation with experiments it could be shown that the gas conversion was sufficiently calculated and the conversion state of the oxygen carrier tracked correctly. The attrition modeling, however, with its parameters measured had to be adapted, so that it fitted the outlet particle flows measured under reaction conditions.

The financial support of DFG (Deutsche Forschungsgemeinschaft) within the priority program SPP 1679: â€œDynamic simulation of interconnected solids processesâ€ is gratefully acknowledged; grant number HA 6935/2-1.

514599 Analysis of FCC Cyclone Fault Diagnosis Technology Based on Particles Information

Abstract:

Cyclones are key equipment in fluid catalytic cracking unit (FCCU), which are used for separating the catalyst particles from oil gas or flue gas. Hence, it is very necessary to carry out condition monitoring and fault diagnosis for the running cyclone. The properties of particles vary at different positions in the particle circulation loop, such as size distribution, morphology and element composition, etc. In consequence, the particles contain much information about operation. Once the cyclone operates abnormally, the information of run-off catalyst particles appear abnormal too, and there exists a corresponding relationship between them. So, taking the catalyst particles as the carrier of fault information, this paper put forward cyclone fault diagnosis technology by extracting catalyst particle information, such as particle packing density, particle size distribution, particle micrographs etc, to build the relationship model between the cyclone fault and catalyst particle information.

514600 The Multi-Hole Throttling Effect and the Erosion Characteristics of the High Pressure Natural Gas

Abstract:

Based on Euler-Lagrange multi-phase flow numerical simulation method, the single hole and multi-hole throttling process of the high pressure natural gas was studied by using the RNG *k-Îµ* turbulent model, the Sutherland Viscosity Law compressible fluid viscosity model, and the advanced erosion model numerical simulation. The calculated mass flow variation law in the throttle choke is comparable with the theoretical calculations, which verified the adopted models and calculating methods. And then the throttling effect and the erosion characteristic of the high pressure natural gas was studied to check its specific law with different throttle holes by using the advanced erosion model. The results showed that, high pressure natural gas can reduce pressure and expansion after flowing through the single hole and multi-hole throttle choke, in which lead a high-speed jet region came up near its outlet. But the maximum jet velocity of multi-hole throttle choke was significantly lower than that of the single hole. At the same time the multi-hole uniform jet may reduce the vortex entrainment and the wall erosion damage, in which can be easier to mix the cold core and achieve stable pressure recovery at the role of the multi-jet mixing effect. Therefore, the multi-hole throttling can effectively restrain the erosion damage and the gas hydrate formation probability.

514601 Numerical Simulation of Flow Field In a Gas Pipe Distributor of the FCCU Regenerator

Abstract:

The flow field in the pipe distributor used in regenerator of FCCU (Fluid Catalytic Cracking Unit), was analyzed based on the numerical simulation. Firstly, the distributing gas performance of the pipe distributor was investigated. The result shows that the pressure increases along the branch pipe of the pipe distributor and the axial flow rate decreases in the branch pipe, which leads to the nozzle flow rate increasing gradually. Furthermore, an obvious bias flow was found in the inlet of branch pipe, which brings some nozzle flow rate less than the designed flow rate and causes an uneven gas distribution in the fluidized bed as well. Meanwhile, the momentum equation with variable mass flow was applied to analyze the flow in the pipe distributor. The flow type in the branch pipe belongs to the "momentum exchange controlling model". The pressure in the branch pipe always increases from the inlet to the end, and the pressure distribution is not consequentially uniform. The uneven gas distribution of the pipe distributor results in fluidized gas non-uniformity in the bed and erosion in the branch pipe and nozzle. Finally, according to the pressure characteristics in fluidized bed system, relationship between the pressure drop of branch pipe and pressure drop of nozzles, as well as of the critical nozzle pressure drop and the designed nozzle pressure drop, were discussed. The gas distributor performance could be optimized by means of the structural improvement of branch pipe.

514602 Full-Loop Simulation of Gas-solids Flow in Circulating Fluidized Bed of FCC System

Abstract:

Circulating fluidized beds are widely used in petrochemical, coal, metallurgy, and other industrial fields. A typical circulating fluidized bed (CFB) includes the riser, gas-solid separator, downer and solids flow control device. To obtain detailed information and get better understand of gas-solids flow in CFB, computational fluid dynamics (CFD) had been applied extensively in many systems such as circulating fluidized bed combustion (CFBC) systems and fluid catalytic cracking (FCC) systems. In most studies, the riser, the core part in CFB system, was isolated as a single simulation object by applying proper boundary conditions. However, there are some limitations of riser-only simulation, without consideration of dynamics in solids circulation rate and effects of other part of CFB system. Therefore, the full-loop simulation is a more promising way to describe complex gas-solids interaction in CFB by treating the whole CFB system as a simulation object. Nowadays, most of current full-loop simulations were focused on CFBC systems, there was little work about FCC system, with different type of particles and recycle mechanism from CFBC system.

In this work, both full-loop and riser-only simulations were carried out by TFM approach to investigate a pilot-scale gas-solid circulating fluidized bed of FCC system which has been set up in our group. The simulation condition was chosen according to experiments with different superficial gas velocity, valve opening and fluidizing air flow rate. Detailed comparison between full-loop and riser-only numerical simulations has been conducted with respect to the hydrodynamics inside the riser, which showed full-loop simulation had advantages on predicting dynamics of solids circulation rate and the gas-solids flow behavior. Distribution of pressure, solids holdup and velocity in system was investigated which was helpful to understand the hydrodynamics in CFB. The effects of operating condition were then studied, which revealed the system stability mechanism.

506758 Immobilization of Sulfur-Oxidizing Bacterium, *Thioalkalivibrio Sp.* D301 on Magnetic Nanaoparticles and Biodesulfurization

Abstract:

There is an urgent need to develop desulfurization methods for halo-alkaline wastewater. *Thialkalivibrio sp.* D301, an obligatory chemolithoautotrophic, haloalkaliphilic, sulfur-oxidizing bacterium, was used in the removal of sulfide in wastewater. We developed a process to immobilize D301 cells by using superparamagnetic Fe_3O_4 nanoparticles (NPs) under halo-alkaline conditions. Fe_3O_4 NPs were synthesized by coprecipitation of ferric and ferrous salts in the presence of N_2 gas. TEM image shows the spherical shape of Fe_3O_4 NPs with an average diameter of 10 to 15 nm. The particles had superparamagnetic properties. The saturation magnetization (Î´$_s$) of the Fe_3O_4 NPs was 73.985 emu gâˆ'1 with zero remanence and coercivity. Analysis of zeta potentials and hydraulic diameters showed that the Fe_3O_4 NPs and bacterial cells interact mainly through the nano-size effect. The NPs-coated cells have the same sulfur oxidization activity as that of free cells. Experiment results showed that the immobilized cells still retained >90% of their sulfur oxidization activity after six times of reuse. The cells coated with Fe_3O_4 NPs did not interact through electrostatic attraction. The mechanism of NPs adsorption mainly involved the nano-size effect. Magnetic immobilization is a convenient technique for HA-SOB immobilization under halo-alkaline conditions and is a promising technology for large-scale industrial applications.

507944 Simulation of Bulk Solids and Granular Systems By Using Combined Discrete Element Models

Abstract:

Granular materials can be found in many industrial processes from chemical to energy, agriculture, mining, pharmacy and many other fields. They are almost always opaque, making the acquisition of experimental data below the surface difficult. DEM (Discrete Element Method) simulations are an effective way to study the physics of granular materials because they can simulate the micro-dynamic behavior of granular material on an individual particle scale and the motion of every particle can be traced. Since DEM was first considered, many modified discrete element models have been developed. However, most existing DEM simulation are just for spherical particles or a specific type of non-spherical particles. In this work, a combination of several types of discrete element models including spherical models, super-ellipsoids, multi-sphere models (with no interaction and deformation between sub-spheres), and bonded-spheres models (with interaction and deformation between sub-spheres) were used to simulate the flow of granular materials with many different shapes. Several simulation cases including the flow of particles in tumblers, blenders, and other devices were performed to validate the modeling approaches. The simulation results show that it is practical and efficient to combine different discrete element models together for simulating both spherical and non-spherical particles in bulk solids and granular systems.

515703 Material Optimization of Perovskite Films using High Throughput Synthesis and Multi-Dimensional Analysis

Abstract:

Over the past years, Perovskites received special attention in the field of renewable energy as they are one of the prime candidates for energy storage and conversion materials. Perovskites act as the active light harvesting layer of a solar cell and provide excellent performance.[1] A common synthesis strategy is through the liquid phase under inert atmosphere.[2,3] However, this comes with the challenge of a huge parameter space that in principle would have to be first synthesized, then characterized and finally optimized based on the therefrom deduced knowledge.

To overcome these limitations, we follow a fully automated approach, which allows automated high throughput (HT) synthesis and HT characterization of perovskites with high accuracy. As an efficient route to explore the parameter space, we use design of experiments (DoE) to find optimized settings for process parameters such as synthesis and annealing time and temperature, but also material aspects like the molar ratio of the reactants.

For all experiments, a Chemspeed© technologies (CST) swing XL platform with a liquid handling tool and a heatable shaking rack was used. The robotic platform is enclosed within an Mbraun© glovebox with inert atmosphere (Nitrogen). After precursor preparation, deposition was realized by a liquid handling tool on glass substrates inserted in a customized aluminum 24 well plate, which is then transferred onto a hot plate for film formation.

In a first step, after annealing at different temperatures, a characterization cascade was developed and applied that contains the subsequent elimination of samples that exhibit undesired properties and side products. Fast absorption analysis, photoluminescence (PL) measurements and finally X-ray diffraction (XRD) were performed to identify optimum chemistry and process parameters. Although this is a very reliable procedure, it is still comparatively demanding in terms of the time effort. Therefore a more efficient, purely optical approach by which the experimental parameter space is only explored in terms of rapid emission analysis was developed in a second step. Noteworthy, both strategies led to the same optimum conditions, however, at significantly reduced time effort.

In conclusion, our work shows how by the combination of automation, DoE and efficient characterization, optimum conditions for new materials like perovskites are rapidly identified. This opens the door towards faster material development which is important whenever new perovskite structures, e.g. lead free materials for industrial applications, are strived for.

Literature

1. Green, M.A.; Ho-Bailie, A.; Snaith, H.J., "The emergence of perovskite solar cells", Nature Photonics 2014, 8, (7), 506-514.
2. Baikie, T.; Fang, Y.; Kadro, J.M.; Schreyer, M.; Wei, F.; Mhaisalkar, S.G.; Graetzel, M.; White, T.J., "Synthesis and crystal chemistry of the hybrid perovskite (CH 3NH3)PbI3 for solid-state sensitised solar cell applications", J. Mater. Chem. A 2013,1 , 5628-5641.
3. Yan, K.; Long, M.; Zhang, T.; Wei, Z.; Chen, H.; Yang, S.; Xu, J., "Hybrid Halide Perovskite Solar Cell Precursors: Colloidal Chemistry and Coordination Engineering behind Device Processing for High Efficiency", J. Am. Chem. Soc. 2015, 137, (13), 4460–4468.

509509 Solid Liquid Separation Via Particle Flow Instability

Abstract:

A new separation strategy using novel particle-flow-instability physics is successfully developed for a *difficult-to-separate suspension* in which there is some combination of a small density difference between solid and liquid, high viscosity, and small-sized particles. The method we propose here requires no dilution for medium, high or extremely high viscosity slurries and can produce effective solidâ€"liquid separation with low capital and operational inputs. The results have relevance for many biotechnological and/or chemical processes.

509543 Milling and Grindability Assessment of Pharmaceutical Materials

Abstract:

Milling is one of the most important unit operations in the pharmaceutical industries. Most of the Active Pharmaceutical Ingredients (APIs) and excipients are produced by crystallisation, and grown to a sufficiently large particle size suitable for isolation and drying. Therefore most of the materials have to be milled to desirable particle sizes. Milling is influenced by a combined effect of milling dynamics (input power, mill geometry, etc) and material properties. Considerable research has been carried out on analysing the particle behaviour in various mills. Nevertheless, the prediction of the milling rate and milled product properties (e.g. product size distribution) is still very challenging.

In this work, the milling behaviour of several pharmaceutical materials in the PicoPlex mill (pin mill unit of the Picoline, Hosokawa Alpine) has been studied. A combination of experiments and Distinct Element Method (DEM) simulations is used to explore the dynamic behaviours of the materials in the mill. The results are then used to validate a novel approach of grindability assessment of pharmaceutical materials using the Scirocco disperser of the Malvern Mastersizer 2000 (Bonakdar et al. 2016). Based on this method the ease of grindability of the materials in the mill can be predicted.

Bonakdar, T., Ali, M., Dogbe, S., Ghadiri, M., and Tinke, A. (2016). A method for grindability testing using the scirocco disperser. *International Journal of Pharmaceutics,* 501: 65-74

507857 Bi-Directional Thermal Control of Twin Screw Granulation Process Via a Specialised Annular Heat Pipe

Abstract:

The continuous process of twin screw granulation forms a hotbed in pharmaceutical research as it is expected to succeed the current batch production of granules. In essence, the scale up of the material output is a function of production time, instead of equipment size (for batch granulators). This way, a â€œjust-in-time productionâ€ of the desired amount of material can be performed according to specific needs. In this work, a twin screw granulator was utilised in the granulation of both the placebo and active pharmaceutical formulations. Subsequently, the same equipment was used in granulating ceramic raw material powders for tile pressing. During processing, the spatial constraint in the barrel and screw compartments caused friction, resulting in excess heat generation. For the thermally-sensitive pharmaceutical formulations, this must be controlled to avoid denaturation of the granule products. A common solution is by using flowing cool water, but a good control has been proven to be difficult to achieve as different granulation zones generate different heat levels (kneading blocks > conveyer blocks). Heat pipe is a passive and highly effective thermal conductor and stabiliser. A carefully-designed annular heat pipe, filled with a small amount of methanol was successfully employed to stabilise the temperature in the barrel space, cooling the granulating material before it reached its critical temperature. For the ceramic powders, the methanol filling of the heat pipe was replaced by water and it was then heated up to provide an in-situ drying operation as the powders granulating. The in-situ drying in this instance, reduces energy consumption for the downstream drying processing as the product granules contain less moisture to be removed. The annular heat pipe, with its bi-directional operation of heating and cooling is a technically viable temperature controller for twin screw granulators. Compared to the established active cooling methods, the heat pipe provides a passive alternative, with only a discreet amount of fluid required to last its operational lifetime, thereby providing an economic thermal control solution.

Mixing Grains with Different Elongation in a Rotating Drums

Claudia Piacenza[1,2], *Marco Marconati*[1], *Colin Hare*[1], *Andrea C. Santomaso*[2], *Marco Ramaioli*[1,*]

[1] *Department of Chemical and Process Engineering, Faculty of Engineering and Physical Sciences, University of Surrey, Guildford GU2 7XH, UK*

[2] *APT Lab, Department of Industrial Engineering, University of Padova, Padova, Italy*

[*] *Email: m.ramaioli@surrey.ac.uk*

Abstract

Mixing grains with different properties is a remarkably challenging process, relevant to many industrial applications. Rotating drums have been used extensively as model systems to study granular media flow and mixing and segregation. Numerous studies considered the mixing of grains with different sizes, but only few studies considered shape and elongation, which have been already identified as important characteristics affecting packing and inducing segregation. In this contribution, the mixing of binary mixtures of grains having the same volume, but different elongations was studied experimentally. The mixing dynamics of a layered granular medium was characterized in a rotating drum, highlighting the impact of the grain shape, of the drum angular speed and of the drum filling ratio. A mixed or segregated state is reached very rapidly, but experiments are continued to verify that the state reached is a steady state. The experiments demonstrated that below a critical difference of elongation, the grains can be mixed effectively. Conversely, when grain elongations are very different, a central core is formed, rich in the more elongated grains. In this case only limited mixing can be achieved. These results can guide the formulation of mixtures of grains with different elongations.

1. Introduction

Mixing is a process operation aimed at obtaining a homogeneous system starting from two or more components. On the contrary, when there is a lack of homogeneity within the mixture segregation occurs. Segregation phenomena represent a relevant problem in the case of mixing solid particles having different physical properties such as particle size, density, shape or surface roughness. Differently from liquids and gases, solid particles are not thermal activated system [1] and so in order to obtain a certain degree of mixing they need to be mechanically activated. Typically three mixing mechanisms are largely recognized:

- dispersion, analogous to diffusion in fluids, it involves single particles that randomly move due to collision between each other;
- convection, it involves a portion of the bulk material, with a given direction and velocity;
- shear mixing: it is peculiar to granular solids and it is a combination of the two previous mechanisms.

Powder mixers are relatively simple machines of low capital cost. They are classified into two types: tumbling mixers and agitated mixers. Tumbling mixers are the simplest. A totally enclosed vessel is rotated about an axis causing the particles within the mixer to tumble over each other on the mixture surface. Tumbling vessels provide low-shear environments. Convective mixers instead are characterized by the action of impellers or paddles to move the powders around and generate a well-mixed product. Convective mixers provide more shear into the mixture and can be used with cohesive materials.

Rotating drums belong to the tumbling mixers. They are extensively used in the chemical and process industries. With respect to mixing processes, rotating drums are often used to study and understand mixing and segregation phenomena; as a result, many experimental and numerical studies were developed on granular behavior in rotating drums [1-3]. Different granular flow regimes can occur in the drum, each one with its specific flow behavior. In particular, different flow regimes can be assessed by using the dimensionless Froude number, $Fr = \omega^2 R/g$ [4], where ω [s^{-1}] is the drum rotation speed, R [m] the drum radius and g [m/s^2] the gravitational acceleration. With $10^{-4}<Fr<10^{-2}$ and filling level $\varphi>0.1$ rolling regime takes place. This motion is characterized by a uniform, steady flow of a particle layer on the surface (cascading or active layer) while the larger part of the bed (plug flow region) is moved upwards by solid rotation with the rotational speed of the wall. A uniform and good mixing is achievable with this type of motion, but when particle size distribution is very broad segregation can appear. The rolling bed is preferred, since is the simplest regime providing favorable conditions for heat transfer and high quality of the product even when mass flow rates are large.

When particles to be mixed have different properties such as different size, different density or different shape segregation typically occurs and particles tend to separate. [5, 6]. Difference in size is by far the most important cause of particle segregation. Particle size distribution influences also strongly bulk properties such as packing fraction and bulk modulus [7].

Different size segregation mechanisms can be identified:

- trajectory mechanism: according to the Stokes' law the velocity of fine particles is lower than that of big particles. This can cause segregation when small particles move through air;
- percolation mechanism: if a bulk material is disturbed, a rearrangement in the particle packing occurs and for fine particles it is easier to fall down in the inter-particle voids;
- inertial mechanism: in pouring and filling operations bigger particles tend to flow selectively to the bottom of the pile since they have a higher momentum;
- elutriation mechanism: in presence of air the very fine particles can remain in suspension while larger ones tends to settled. This can occur in charging storage vessels or hoppers.

A homogeneous binary granular mixture, with differences in size and density, in a two-dimensional rotating drum typically gives radial segregation. Considering a rotating drum operating in the rolling regime filled with particles of different sizes, bigger particles mainly tend to the edge of the system, while smaller particles accumulate to the centre of the drum. The phenomenon is fast and the rotational speed of the drum seems not to affect the phenomenon; the size ratio of the particles is, instead, the parameter that mainly affects size segregation phenomena. Most of the experimental investigations about segregation in the rotating drum involved spherical or nearly-spherical particles; the experimental studies of the flow of non-spherical particles are, in fact, rather limited [6,8]. The shape of the particles, however, strongly influences the motion in the system. In particular, the static and dynamic angles of repose of non-spherical particles are found to be very different from the ones measured for spherical particles suggesting that differences in particle mobility can affect the segregation process [9].

2. Materials and methods

2.1 Particles

Both spherical and non-spherical particles were used for the experiments. Their size (diameter d for spheres, major and minor axis for ellipsoids a and b respectively) , aspect ratio AR, projected area A_p, volume V, and true density ρ are reported in Table 1.

Table 1. Physical properties of the particles

Type	d [mm]	a [mm]	b [mm]	AR [-]	A_p [mm^2]	V [mm^3]	ρ [g/cm^3]
Spheres S	8	-	-	1	50.3	268	1.09
Ellipsoids E1	-	10.48	6.98	1.5	57.5	267	0.96
Ellipsoids E2	-	18.44	5.26	3.5	76.2	267	0.98

Spherical particles with size 8 mm were used for the experiments. The ellipsoids were 3D printed with *AR* equal to 1.5 and 3.5. They were printed with the Ultimaker 3D printer and made of polylactic acid (PLA). The software used to realize the 3D design of the particles used in this work is FreeCAD, which is an open-source parametric 3D modeller allowing to obtain STL files.

The true density of all the particles used for the experiments was measured by gas pycnometry (Multivolume Pycnometer 1305, Micromeritics). The measured density of the spheres was very close to the typical density of the nylon 66 (1.14 g/cm^3). The PLA ellipsoids had a density that was lower than the typical density of the PLA (1.24 g/cm^3). This is because the particles were printed with a set infill density of the 20%, so they where partially empty inside.

2.2 Tumbling mixer

The quasi-2D drum was made of two transparent polycarbonate disks with a diameter of 30 cm and a thickness of 6 mm; between them a wooden spacer was placed, with the same external diameter, an internal diameter of 25 cm and a thickness of about 9.1 mm. The spacer allowed to create an internal volume of finite thickness separating the two polycarbonate disks.

a) b)

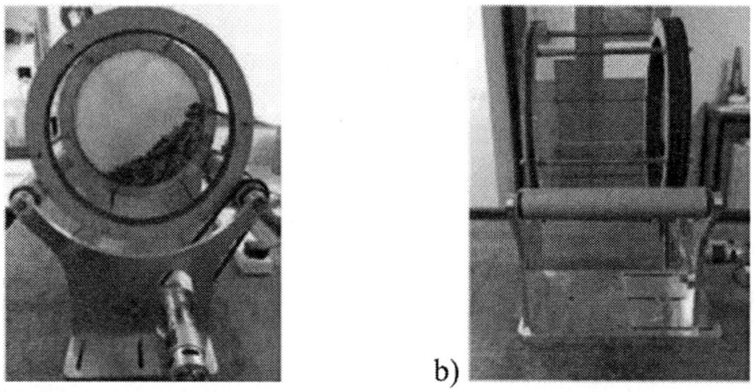

Figure 1. a) Front view and b) side view of the rotating drum (the drum is on the left side of the rigid structure).

The internal lateral wall of the volume was covered with sandpaper in order to provide friction and avoid undesired slip of the particles. To keep the drum in vertical position and allow its rotation, three additional wooden spacers were used, at a distance of about 16.3 cm. This rigid structure could rotate at variable rotational speed thanks to a motor connected to a voltage supplier. Two rollers mounted on a rigid support made the rotation of the drum possible; a pulley transmitted the motion to the driving roller, while the second roller was passive. The rotational speed could be changed by varying the motor voltage. The structure could be removed from the rollers and by unscrewing the bolts, the drum could be easily opened so that to load/unload the particles.

2.3 Image acquisition

Frontal pictures of the drum were taken by a color camera and analyzed by the image processing software ImageJ [10]. To avoid the reflection of the camera and other objects on the front transparent disk, a squared LED panel was used which allowed to have a dominant diffused light illuminating the drum. The LED had a power of 8 W and a color temperature of 4000 K; it was placed behind the drum and opposite to the camera. A circular sheet of paper, which was attached to the drum between it and the LED, diffused the light properly. The camera was a Basler ace camera – model acA2000-165uc, with a maximum resolution of 2040 × 1086 pixels. The focus and the diaphragm of the camera could be adjusted manually.

2.4 Experimental plan

The attention was focused on mixtures of particles with different shapes, since segregation phenomena due to size had been widely investigated in the past. Three mixtures were considered, with two possible initial top-bottom segregated configurations. Filling level and rotational speed where changed also as detailed in Table 2.

Table 2. Experimental tests performed with differently shaped particles.

AR 1 [-]	*AR* 2 [-]	ω [rpm]	φ [-]	Init. Config.
1	1.5	4.22	25%	E1/S
1	1.5	4.22	25%	S/E1
1	1.5	4.22	47%	E1/S
1	1.5	4.22	47%	S/E1
1	1.5	11.51	25%	E1/S
1	1.5	11.51	47%	E1/S
1	3.5	4.22	25%	E2/S
1	3.5	4.22	25%	S/E2
1	3.5	4.22	47%	E2/S
1	3.5	11.51	25%	E2/S
1	3.5	11.51	25%	S/E2
1	3.5	11.51	47%	E2/S
1.5	3.5	4.22	25%	E2/E1
1.5	3.5	4.22	25%	E1/E2
1.5	3.5	4.22	47%	E2/E1
1.5	3.5	4.22	47%	E1/E2

*AR*1 and *AR*2 indicate the aspect ratios of the particles in the binary mixture. It is reminded that the spheres (*AR*=1) and the two types of ellipsoids (*AR*=1.5 and *AR*=3.5) had the same volume. The E1/S initial configuration indicates a segregated configuration with short ellipsoids above spheres (this was assumed as basis configuration); the S/E1 configuration is inverted with spheres above short ellipsoids. A similar convention is applied to the experiments considering mixtures of ellipsoids. All the experiments were performed with binary mixtures with a composition of 50% and 50% on a number (and hence also volume) basis. The filling level φ of the 25% was obtained with a total number of particles equal to 200; with 320 total particles the filling level of the drum was equal to 47%.

2.5 Image analysis procedure

Image analysis techniques were used to identify the particles on the basis of their different shape. The procedure was automated using customized ImageJ macros. Contrast and brightness of the images were first adjusted. Color images were then set to gray scale and segmented by applying a threshold. The operation was carried out twice to separate light and dark particles with different thresholds. Watershed segmentation was then used to separate touching objects. This last operation was more difficult for the particles at the border of the drum; in some case, in fact, the ellipsoids were erroneously identified as spheres. Also, the presence of the border of the drum complicated the watershed operation.

The centre of mass of each particle ($x_{m,i}$ and $y_{m,i}$) was estimated and the radial distance (d_i) from the centre of the drum (x_c, y_c) calculated as:

$$d_i=[(x_c-x_{m,i})^2+(y_c-y_{m,i})^2]^{0.5}$$

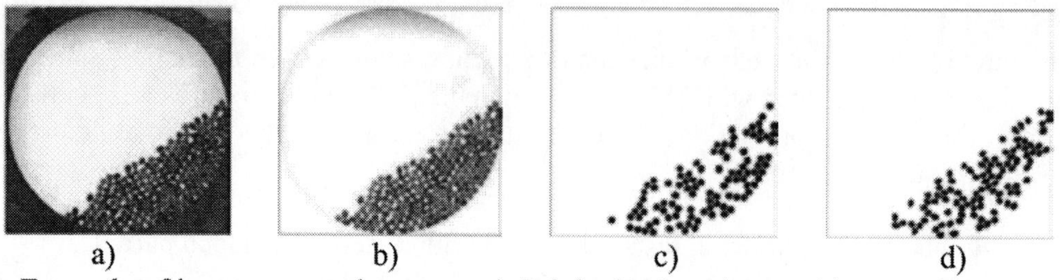

a) b) c) d)

Figure 2. Example of image processing steps. a) Original colour image; b) grayscale conversion and circle cut from the original picture; segmentation applied to c) light particles and d) dark particles.

The mean value of distance (d_m) for the spheres and for the ellipsoids was then calculated as:

$$d_m = \Sigma\, d_i/N$$

Only for very elongated ellipsoid (AR=3.5) the watershed operation failed resulting in over-segmentation errors. Single particles could not be identified unequivocally in this case; only one center of mass for all the elongated ellipsoid was estimated and the average distance (d_m) of the particles from the drum center calculated.

3. Results and discussion

Figure 3 shows two typical experiments, carried out at the lowest rotational speed and filling ratio, with ellipsoids (AR=1.5 at left and AR=3.5 at right) on top and spheres on bottom in the initial and final configuration. It is evident even by visual inspection that the final state presented some level of radial segregation and that a different final segregation intensity was reached in the two systems.

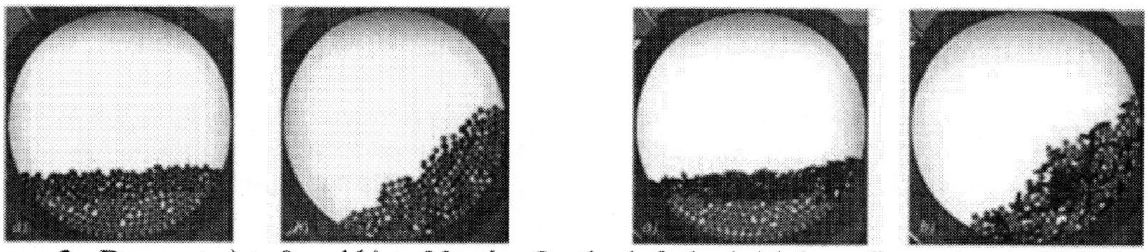

Figure 3. Drum at a) t=0 and b) t=30 min. On the left the initial configuration is E1/S (using ellipsoids with AR=1.5) and on the right the initial configuration is E2/S (AR=3.5). In both experiments ω=4.22 rpm and φ= 25%.

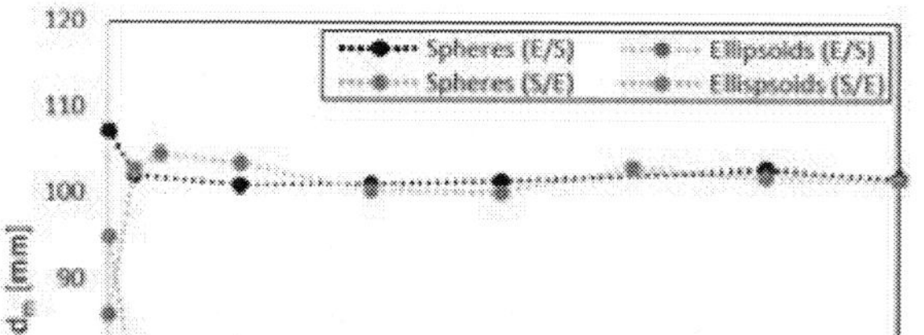

Figure 4. Evolution of the mean radial distance vs. time for the mixture of spheres and ellipsoids with AR=3.5. The initial configuration E2/S is compared to the initial configuration S/E2.

A more quantitative analysis was obtained by plotting the trend of the mean radial distance (d_m) of the two species at different time instants. As a typical example of the system evolution, the case of E2/S

is shown in Figure 4. The d_m reached a steady value after a short time. Dedicated experiments on a shorter time scale (2 minutes, not shown here) revealed that steady state value for d_m was reached after 30s for mixtures with ellipsoids E1 (AR=1.5) and after 60s for mixtures with E2 (AR=3.5).
At steady state, the radial distance was in both cases smaller for the ellipsoids, meaning that they mainly remained concentrated on top or at the center of the granular bed while the spheres mainly concentrated at the periphery of the drum (close to the drum wall). This tendency was much stronger for ellipsoids E2. So a slower segregation dynamics compensated by a larger segregation intensity was observed with the more elongated particles.
As Figure 4 shows, inverting the initial configuration from E/S to S/E did not affect the dynamics and the final result in all the experiments performed with these particles. Initial configuration resulted to have no effect on the segregation propensity of the mixtures analyzed.

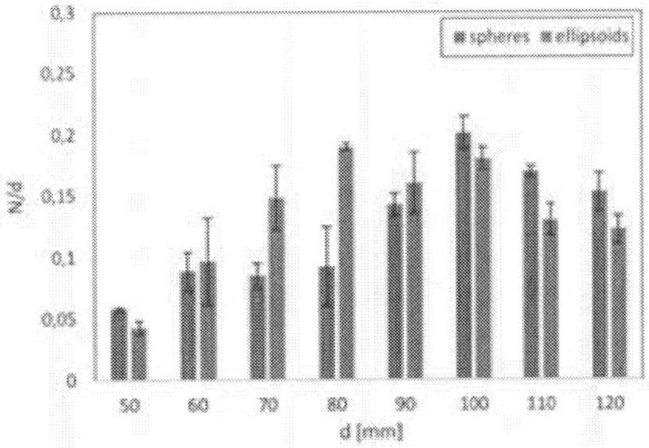

Figure 5. Distribution of the radial distance of spheres and ellipsoids with AR=1.5, obtained by averaging the results of three repetitions. Other conditions are t=30 min, ω=4.22 rpm, φ= 25%.

A more in-depth analysis can be carried out considering the distribution of the radial distance of the spheres and ellipsoids separately. The radial distance from the centre of the drum (d) was divided in classes of 10 mm and the number of each type of particles (N) in each class was computed. To obtain the distribution, N was normalized on the radial distance as N/d and plotted versus d. In the case of completely homogeneous mixtures and assuming that the medium takes the shape of a circular sector, N/d should be constant and equal for the two types of particles in each class. The peak of the distribution of the ellipsoids was found at lower distance than the peak of the distribution of the spheres. Moreover, at lower distance, the number of ellipsoids was higher than the number of spheres, especially for distances between 70 and 90 mm; at higher distance (closer to the drum wall), instead, the number of spheres was higher than the number of ellipsoids, thus confirming the previous analysis based on d_m.

Also, binary mixtures made of the two types of ellipsoids E1 and E2 (AR=1.5 and AR=3.5 respectively) were taken into account at the same filling levels of the previous experimental tests and at the lower rotational speed. Stable segregation was again observed with intermediate results with respect to the two previous cases (S+E1 and S+E2). So as a general comment on these experiments, stable segregation was always observed when mixing particles having same volume but different AR. The intensity of segregation increased with increasing the difference in AR between the particles. In particular particles with larger AR tended to accumulate at the center of the granular bed, while particle with lower AR accumulated at the periphery. Segregation occurred in the surface active layer where the more elongated particles, because of their reduced propensity to flow (particle rolling was prevented by their elongated shape), were systematically overtaken by the less elongated (and faster) particles. In this way a spontaneous separation of particles occurred in the active layer, with those having a lower AR arriving first at the drum wall, while those having an higher AR remaining trapped in the core.

3.1 Effect of filling level

The comparison of the behavior of the mixtures at the two filling levels can be carried out by normalizing the mean radial distance d_m of the two types of particle on the mean distance of all the particles d_c, calculated as $(d_{m,\ spheres}+d_{m,\ ellips.})/2$. The trends of d_m/d_c are shown in Figure 6

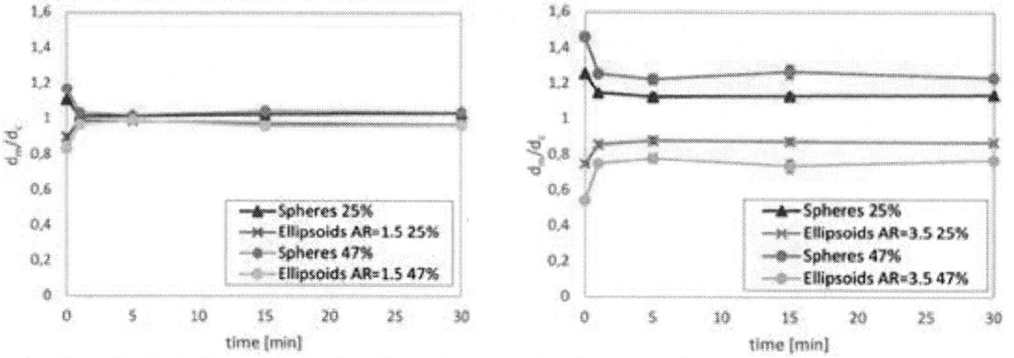

Figure 6. Trend of d_m/d_c vs. time for the mixture of spheres and ellipsoids with *AR*=1.5 (left) and AR=3.5 (right) at two different filling levels.

In the case of mixture of spheres and ellipsoids with *AR*=1.5, the curves at different filling levels overlap, showing that the behaviour of the mixture was not affected by the filling level of the drum. In the case of the other two mixtures (S+E2 and E1+E2, not shown), the curves did not overlap , especially for the case S+E2 shown in Figure 6 (right). An explanation can be found in terms of bed geometry. When increasing the filling level (below 50%) the length of the surface active layer increased as well. This gave more chances for particles to separate, in particular when the difference in *AR* was large enough (S+E2 and E1+E2). However for the system S+E1 where the difference in *AR* was the smallest, the difference in active layer length in the two cases was not sufficient to create observable differences in the segregation pattern.

3.2 Effect of rotational speed

In addition to experiments carried out at 4.2 rpm, tests were repeated at 11.5 rpm with spheres and ellipsoids (25% filling level). Figure 7 shows that rotational speed did not affect the segregation patterns, at least in the range of the analyzed speeds.

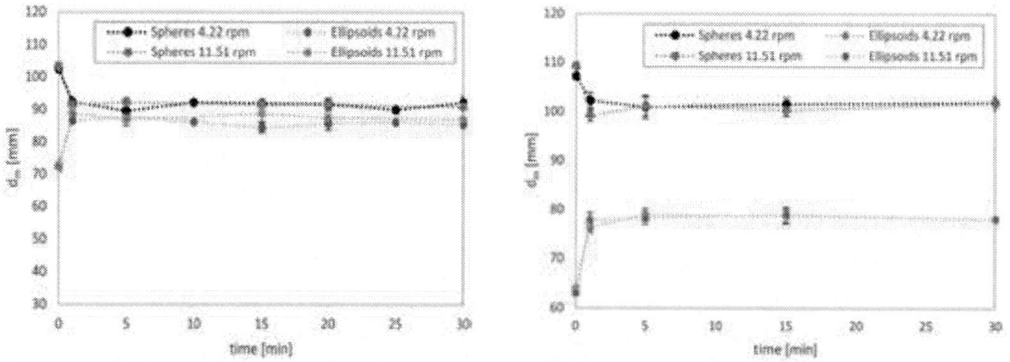

Figure 7. Trend of the mean radial distance vs. time for the mixture of spheres and ellipsoids with *AR*=1.5 (left) and *AR*=3.5 (right) at two different rotational speeds.

Conclusions

From the experimental tests described in the previous paragraphs, it was found that binary mixtures of particles with different shape but with the same volume do not mix uniformly and segregation was always observed. In particular, the particles with lower aspect ratio tended to stay closer to the drum walls, while the more elongated particles were found preferentially at the centre of the granular bed. The most significant case was represented by the mixture of spheres and ellipsoids with AR=3.5; for this mixture, in fact, the segregation was more evident and the non-spherical particles clearly accumulated in the centre of the granular bed. The same final state was obtained regardless of whether the longer particles were initially located to the top or the the bottom of the mixture. A proportionality between AR difference and segregation intensity was observed. Some variables such as rotational speed and initial bed configuration were found to not affect the segregation dynamics and intensity. Only filling level (in addition to AR) was found to affect segregation intensity. A mechanistic explanation was given in terms of differences in particle flowability occurring in the surface active layer (confirmed also by an analysis of the angles of repose in the drum). In this specific case therefore percolation and inertial effects (typical of size segregation) were not the main driving mechanisms for the observed segregation (the particle indeed had the same volume and similar mass so that sterical and inertial phenomena were minimized). Segregation mainly occurred because of differences in particle velocity in the active layer due to a reduced mobility of particles with larger AR.

References

1. GDR MiDi (2004), "On dense granular flows", The European Physical Journal E, 14, 4, 341–365.
2. Jarray A., Magnanimo V., Ramaioli M., Luding S. (2017) Scaling of wet granular flows in a rotating drum, EPJ Web of Conferences, 140,03078.
3. Volpato S., Canu, P., Santomaso A.C. (2017), "Simulation of free surface granular flows in tumblers", Advanced Powder Technology 28(3), 1028-1037.
4. Mellmann J. (2001), "The transverse motion of solids in rotating cylinders – forms of motion and transition behaviour", Powder Technology, 118, pp 251-270.
5. Windows-Yule C.R.K., Scheper B.J., van der Horn A.J., Hainsworth N., Saunders J., Parker D.J., Thornton A.R. (2016), "Understanding and exploiting competing segregation mechanisms in horizontally rotated granular media", New Journal of Physics, 18, 2, 023013, 2016.
6. Santomaso, A.C., Olivi, M., Canu, P. (2004), "Mechanisms of mixing of granular materials in drum mixers under rolling regime", Chemical Engineering Science, 59 (16), 3269-3280.
7. Kumar N., Magnanimo V., Ramaioli M., Luding S. (2015), "Tuning the bulk properties of bidisperse granular mixtures by small amount of fines", Powder Technology, 293, 94-112.
8. Ramaioli M., Pournin L., Liebling Th. M., (2005), "Numerical and experimental investigation of alignment and segregation of vibrated granular media composed of rods and spheres", Proceedings of the 5th International Conference on Micromechanics of Granular Media, 2, 1359-1363.
9. Santomaso, A.C., Canu, P. (2007), "Single particles properties vs. bulk flowability", Proceedings of 5th International Conference for Conveying and Handling of Particulate Solids.
10. Schneider, C. A., Rasband, W. S., Eliceiri, K. W. (2012), "NIH Image to ImageJ: 25 years of image analysis", Nature methods 9(7): 671-675.

507441 Generation of Particles with a Special Morphology By Desublimation of Copper Phthalocyanine

Abstract:

The morphology of submicron particles becomes more and more important concerning new functional materials. For example the special electric properties of particles can be used in electronic devices [1]. Moreover the electric, optical or catalytic properties of particles can be optimised by adjusting the morphology of the particles. This may imply a risk for the health but makes it possible to realize specific product characteristics. The morphology of the particles should therefore be considered e.g. for filter tests and for the developing and testing of sensor systems.

To explore the characterization of the particle morphology the DFG-AiF cluster project "multi-parametric characterization of particle-based functional materials using innovative online measurement systems" (MPaC) includes the development of online sensor systems for the size and structure of the particles as well as the development of generation methods for particles with a special morphology. The production of spherical particles can be achieved with different commercial aerosol generators. Several generation techniques for special particles are described in literature [2] [3] [4]. To test the sensor systems with particles of different shapes a new type of aerosol generator was developed. The aim for the new generator is to produce the particles in the gaseous phase and provide the particles directly for sensor or filter testing.

The developed generator uses the desublimation of copper phthalocyanine (organic pigment) to structure the surface of titanium dioxide. The bulk material copper phthalocyanine is dispersed into a nitrogen gas stream and transported to a tube furnace. Here the bulk material sublimates and the gas stream carries the material to a laval nozzle where the adiabatic expansion takes place. This leads to a rapid decrease of the temperature behind the laval nozzle and an oversaturation in the gaseous phase. In addition to the decrease of the temperature core particles of titanium dioxide were initiated behind the laval nozzle. The copper phthalocyanine desublimates on the core particles and causes a structured surface. The morphology of the generated particles is characterized by SEM. The particles show a punctual coating of the titanium dioxide particles with copper phthalocyanine. The size of the titanium dioxide particles is about 500 nm and the punctual coating of copper phthalocyanine measures a few nanometres.

The generation method and the design of the generator will be presented and the morphology of the generated particles will be described.

[1] K. Deppert, M. H. Magnusson, S. B. Carlsson, B. Gustafson, T. Junno, . L. Montelius, B. J. Ohlsson, L. Samuelson und L. E. Wernersson, „Future electronic devices by aerosol techniques," *Journal of Aerosol Science,* Nr. 31, 2000.

[2] J. Raula, A. Kuivanen, A. Lähde, H. Jiang, M. Antopolsky, J. Kansikas und E. . I. Kauppinen, „Synthesis of image-leucine nanoparticles via physical vapor deposition at varying saturation conditions," Journal of Aerosol Science, Nr. 38, 2007.

[3] M. Heurlin, M. H. Martin, D. Lindgren, M. Ek, R. L. Wallenberg, K. Deppert und L. Samuelson, „Continuous gas-phase synthesis of nanowires with tunable properties," Nature, Nr. 492, 2012.

[4] K. Nielsch, K. Deppert, M. H. Magnusson, G. Dumpich, E. Kruis und H. Fissan, „Formation of ultrafine particles from powders of compounds with incongruent sublimation behavior," Journal of Aerosol Science, Nr. 28, 1997.

516184 Rubbery Milling of Seed Endosperms for Improved Sustainability by Natural Functionality Preservation

Abstract:

Increasing demand for functionalised goods addressing individual needs requires a profound understanding of process-induced alterations. Development of models and process-structure-property relationships are essential to control the resulting quality and properties of processed food. A wide range of food systems contain solid or semi-solid particulates as major structure building blocks. Generation and tailoring of such particulates strongly depends on their size distribution, shape, morphology, phase behavior (amorphous/crystalline), interfacial properties and related inter-particle force interactions.

A novel processing technology was developed widening the frame of possible alterations, aiming towards tailoring such particle properties generating novel food structures or preserving techno-functional characteristics of the raw material. The technology employs targeted adjustments of the initial moisture content in combination with milling temperature. The resulting transition from the glassy to the rubbery state generates different material properties and processing behaviour. The technology is highly versatile, owing to the universality of moisture induced state transitions.

Seeds from locus bean gum or guar gum were wetted with excess water in order to adjust a rubbery state of the seed endosperm material. In this state milling within a novel type of impact mill with superimposed drying and sifting steps causes more gentle material disintegration compared to conventional dry milling. This preserves the high molecular weight fraction of e.g. galactomannans (micro-scale structure) from degrada-tion and allows at the same time the adjustment of a micro-porous morphology (macro-scale structure) improving wetting, water uptake and dispersing characteristics of the resulting thickener powders. Such structure preservation /modification increased the thickening characteristics extraordinarily as will be demonstrated.

From this prominent rubbery milling example we have entered in related systematic work addressing a wider range of seed endosperms with various â€œnature-synthesizedâ€ func-tionnalities to be further addressed.

/1/ Pollard, M., Eder, B., Fischer, P., and Windhab, E. J. (2010) "Characterization of galacto-mannans isolated from legume endosperms of Caesalpinioideae and Faboideae subfamilies by multidetection aqueous SEC." *Carbohydrate Polymers*, 79(1), 70-84.

/2/ Illmann, S., Stoob, M., Sieber, C., Konz, A.K. and Windhab, E.J. (2012) "New Thickening Quality of Galactomannan Polysaccharides by Tailored Impact Milling" *XVIth Int. Congress on Rheology*, Lisbon, Portugal, August 5-10

503422 Particle Technology Education at Purdue University

Abstract:

Degree programs and coursework at Purdue University focused on particle technology will be presented. Additional information may be found at: https://engineering.purdue.edu/CP3

506453 Teaching Particle Technology in Portugal â€“ University of Coimbra

Abstract:

Education in Particle Technology started in Portugal in the 80â€™s (last years of the decade). The Chemical Engineering degree of the University of Coimbra was the first engineering degree, in the country, to include this topic in the curriculum. Particle Technology was recognized by us, at that time, essential for the education of chemical engineers, in parallel to education in the traditional areas of Chemical Engineering which were still focused on monophasic systems.

Since then, Particle Technology has always been present in the Chemical Engineering curriculum of the University of Coimbra, in different configurations. During the first years particle characterization was combined with design of operations involving particulate systems (grinding and pneumatic and hydraulic transport). From 2005 to 2010, with the curriculum reformulation in relation with the Bologna Process, the two topics started to be addressed in different courses: one course on Materials Characterization and the process topic in a course entitled Structuring and Stabilization Processes, dealing with colloids and emulsions, grinding, agglomeration and spray drying and, for the first time, an integrated modelling strategy for particulate processes based on the population balance formulation was introduced in this course. The two courses were compulsory, at this time.

From the school year 2011/2012 till present, the configuration of this area in the Chemical Engineering degree of the University of Coimbra changed again, even if not deeply (see https://apps.uc.pt/courses/EN/course/361). The course on materials characterization, which includes general notions on particle characterization is kept as an autonomous course included in the 1st cycle of the Chemical Engineering degree of the University of Coimbra. A second course on design of processes involving particulate systems now exists in the 4th year of the MSc degree (1st year of the 2nd cycle), entitled Disperse Systems Technologies, which is an optional course, and which deals with all types of disperse systems, i.e gas, liquid or solids as disperse phase. This course starts with a review of the mechanisms/interactions occurring at the interface with the disperse phase, goes on, again, with a chapter on population balances modelling, and proceeds with strategies for disperse systems stabilization, linking this subject with formulation concepts, ending with the design of some processes involving particulate systems, such as colloids and emulsions, crystallization, grinding, agglomeration and spray drying. This course is also offered to other engineering degrees, such as Materials Engineering and Biomedical Engineering.

Also at the PhD level a course on Disperse Systems Technologies can also be offered during the formal lectures period of the PhD program on Chemical Engineering (or others), depending on the students background and interests.

Since the introduction of the Particle Technology topic in the curriculum of the Chemical Engineering degree of the University of Coimbra, the feedback from employers has been very positive, and the competences in this field acquired by our students have been quite valued by the job market. Till today, the Chemical Engineering degree of the University of Coimbra is still the one, in Portugal, with a stronger emphasis on this topic in the curriculum. In our curriculum, the topic of Particle Technology interacts directly with another course on Product Design, also in the 4th year of the MSc degree and, of course, also with the Project Design course (5th year of the MSc degree), besides other more traditional courses in a Chemical Engineering degree such as in Reaction Engineering and with courses related to environmental technologies.

At present, we have started developing multimedia tools for the teaching/learning of this topic, especially in the field of particle characterization, to be made available on-line, as is the case of the interactive video on nanoparticles characterization (http://labvirtual.eq.uc.pt/siteJoomla/index.php?option=com_content&task=view&id=277&Itemid=470, in Portuguese). Also, the European Federation of Chemical Engineering (EFCE) is considering investing in this type of tools for the field of particle technology, as the result of a collaboration between the Working Groups of Characterization of Particulate Systems and Education.

Acknowledgement

The authors would like to acknowledge the financial support of Pest/C/EQB/UI0102/2013 from FCT/MCTES (PIDDAC), co-financed by the European Regional Development Fund (ERDF) through the program COMPETE (POFC).

506629 Highlights of Particle Technology Teaching in Singapore

Abstract:

Our approach to education in particle technology is to incorporate elements of our research in this field, relevant industrial case studies and apply a Technology Enhanced Learning (TEL) approach for implementing pedagogical interventions to enhance the learning experiences of students.

For supercritical CO_2 processing, case studies in extraction and drying are used to teach different chapters in the curriculum such as:

- Particle characterization
  - The bulk density, size, shape of real industry samples are given to students for analysis using software such as ImageJ
- Flow through packed bed
  - The pressure drop and scale-up of packed bed design for supercritical CO_2 flow through packed bed are used as a worked example and assignment problem to help students appreciate the role of density and viscosity in the pressure drop in a packed bed
- Drying
  - Supercritical CO_2 as an alternative drying media as compared to hot air or superheated steam. Case studies and examples provided

For gasification and chemical looping combustion applications, case studies in the treatment of wastes and carbon capture are similarly used as part of the syllabus:

- Particle characterization
  - The effect of polydispersity on various fluidization characteristics (entrainment, transport dis-engaging height, clusters, segregation, etc)
  - The effect of agglomeration
- Reliability of correlations and simulations
  - Correlations in literature give predictions orders-of-magnitude apart
- Scale-up
  - Rule-of-thumbs in scaling up the various regimes

A design software has been developed within the paradigm of Technology Enhanced Learning (TEL) to facilitate learning via a repeated practice approach. The software generates detailed solution steps to typical engineering design problems encountered within this module. Students can compare the complete solutions generated with their own hand calculations, and thereby apply a repeated practice approach towards their learning of engineering design calculations. The software has been successful in enhancing the performance of students in the final examination.

507116 Particle Technology Course at the University of Salerno

Abstract:

The Course of Particle technology at the University of Salerno is a 6 EC, that is a 60 hours course including lectures, as well as classroom and laboratory practical. The course dedicated to Master student in Chemical Engineering. It a compulsory course within the â€œinnovative process and nanotechnologies curriculumâ€ can be optioned by any mechanical or chemical engineering student at the master level.

The course includes basic particle size and shape characterization concepts, fundamentals on interparticle forces, fundamentals of powder mechanics and silo discharge, fundamentals on-fluid particle interactions and fluidization.

The main learning outcome is the ability of the student to understand the phenomena occurring in processes involving particulate solids and to carry out basic design operation on storage and fluidized bed units.

507382 Professional Master of Engineering Degree in Particle Technology

Abstract:

Particles are ubiquitous in the process industries yet nearly absent in the undergraduate education of chemical engineers. This poster descibes a professional Master of Engineering degree program at University of Delaware (the MEPT) that closes that gap by exclusively focusing on particulate matter, its processing, and its incorporation into products. The program can be completed in a single year although we welcome already-employed students who wish to take it part time. Now in our fourth year, we will describe our program and our experiences including guiding distance learners successfully through the program.

In its 1-year realization, a core curriculum of 18 credits is augmented by 6 credits of electives which the student may use to tailor the program to their specific interests. The program is capped by a 10-week, 6-credit summer internship. The 12 core credits taken in the fall teach the fundamentals of particle transport, particle properties & characterization, particle formation and transformation kinetics and the mathematics of particulate systems. The 6 core credits taken in the spring teach the student to apply the fundamentals to both (a) equipment design (in a unit operations course) and (b) product and process design (in a product engineering course).

The presenters of this poster are the program directors, both with long industrial careers preceding this venture into education. In addition, we have developed a strong network of industrial partners who share a wealth of additional experience and expertise with our students both as adjunct professors and as internship mentors. We have successfully placed international students in internships at US companies via the curriculum practical training program (CPT) who have gone on to practice in industry after graduation under the optional practical training program (OPT).

507780 Recent Developments in Particle Technology at the Universidad Tecnica Federico Santa Maria

Abstract:

The Department of Mechanical Engineering of the Universidad Técnica Federico Santa María formed in 2003 a Research Center for Transporting Granular Materials (CITRAM) to support educational activities and experimental research, as an extension for those ME students taking the regular course on Bulk Solids Handling and those developing a thesis in this area.

Over the years, many ME students have studied, investigated and modeled different topics including: pneumatic conveying of solid particles of different nature, size and density; identification of flow patterns developed in the pipeline, minimum conveying velocity and pressure drop; abrasive wear using a novel rubber-wheel tester; segregation of ores in stockpiles and silos; flow of granular materials in hoppers and chutes; heat transfer of granular materials; among others. The aim of this poster is to present and share the results of this work.

508520 Using Perusall to Enhance Student Learning of Particle Technology at Graz University of Technology

Abstract:

Perusall is a digital platform that allows students to collaboratively annotate textbooks, papers, and other similar classroom reading assignments. In order to enhance student learning, Perusall was introduced to help teach Particle Technology at Graz University of Technology (TU Graz, Austria). At Graz University of Technology, two Particle Technology courses exist in sequence; the first course is focused on fundamentals topics, whereas the second course is focused more on applications and separations topics. Approximately 60-75 students are registered for each course. For both courses, Perusall was used to assign reading materials, which were primarily journal articles that complemented the course topics. Although these assignments were non-compulsory due to national requirements, students were asked to use Perusall to read and annotate the course readings. Student participation and input regarding this digital approach to learning was recorded. This poster will highlight how Perusall was implemented in both Particle Technology courses as a tool to enhance student learning.

508524 A Graduate Course in Fluidization and Gas-Solid Flow Systems

Abstract:

The objective of this course is: a) to provide students with a fundamental understanding and design of Fluidization and Fluid- Particle flow systems, b) to provide students with an understanding of Computational Fluid Dynamics (CFD) and the kinetic theory approach to fluid-particle flows c) to provide students with the knowledge required to solve flow, heat and mass transfer problems in processes which involve particles and d) to provide students with sufficient knowledge to read and understand technical articles in areas related to fluid-particle processes. The course is designed for Master and PhD students in Chemical, Mechanical and aerospace engineering. The course includes: Fluidization concepts and regimes including bubbling, turbulent and circulating beds; Particle characterization, diagnostics and size reduction and agglomeration/granulation; Concept and design of Standpipes, non- mechanical valves, gas distributors, cyclones and hoppers; CFD approach to fluid-particle flow systems and fluidization including kinetic theory and dispersed phase modeling; Computational transport phenomena of fluid-particle systems including method of moments; and examples from energy, chemical and pharmaceutical industries. In addition the students will visit PSRI to see large-scale fluidized beds in operation.

The students are required to finish a team project. The teams consist of 3-4 students working on a project which will count as their final exam. The projects will primarily involve background research (about 80%) but some of the investigation (about 20%) should involve taking an additional step from what is currently known. This additional step can take the form of a research proposal, new theory, new insight based on current information, new calculations, new simulation and/or new data. The results of the projects will be communicated to the class and to the Course Instructors via a written report of 10-20 pages, and an oral presentation involving the professors and all of the group members.

508742 Education of Fluid-Solid Multiphase Flow at Department of Mechanical Engineering, Osaka University

Abstract:

In mechanical engineering field, education curriculum in undergraduate course concentrates on the basic subjects of mechanical engineering such as mathematics, mechanics, material mechanics, fluid mechanics, thermodynamics, control engineering, design, etc. Concerning the course relating to powder technology, we have a class of "Fluid-Solid Multiphase Flow" in the curriculum of graduate school of engineering.

The contents of "fluid-solid two-phase flow" is as follows;

1 Introduction: examples of multiphase flow

2 Definitions and fundamental concepts

3 Particle response time and Stokes number

4 Fluid force acting on particle

5 Fluid force acting on particle in dense system

6 Particle-wall and particle-particle interaction

7 Heat and mass transfer

8 Continuous phase flow equations

9 Modeling of fluid-solid multiphase flow

10 Discrete particle modeling and simulation

11 Numerical simulation method for continuous phase flow

This course includes basic concepts of powder technology, dynamics of particles in fluid flow, derivation of basic equations for locally-volume-averaged fluid flow, discrete particle modeling including DEM (Discrete Element Method), and basics of numerical analysis and modeling.

509420 Bulk Solids Handling Education at the KSU Bulk Solids Innovation Center

Abstract:

Kansas State University Bulk Solids Innovation Center (BSIC) was started in 2015 with university and industry partnership and is located in Salina, Kansas. This center is dedicated to bulk solids handling research, education and outreach and equipped with world class pneumatic conveying system (both dense and dilute) full scale test lab and other material testing equipment. The vision of this center is to become valued resources for companies who produce or handle bulk solids during their operations. Currently, this center offers short courses in various subject lines related to bulk solids handling such as storage and flow, pneumatic conveying systems, discrete element method, dust explosion and control and mixing and segregation. Duration for these short courses varies from two to three and half days and offered as live course with class room teaching and practical training. The fundamental course on bulk solids handling is offered as distance education online format. This center is also in the process of starting two year online (distance education) master degree program named professional master of technology specialized in bulk solids handling and one year graduate certificate program (online) in bulk solids handling.

507337 A Rational Single Particle Design Approach Using an Acoustic Levitation System and X-Ray Tomography

Abstract:

The understanding and optimisation of particle morphologies is a common objective for particle engineering approaches either to improve material-handling in the manufacturing process or to influence Critical Quality Attributes (CQAs) linked to product performance. This study demonstrates the rational development of pharmaceutical formulations by conducting single particle experiments to investigate the range of achievable particle attributes across a pragmatic design space.

An acoustic levitator was employed to provide a container-less particle design environment for single droplet drying experiments with starting droplet volumes between 10 ul â€" 80 ul. The temperature and humidity was controlled by an enclosure. The system gave access to characterise the particle formation and subsequent drying process using thermal and high speed imaging techniques. The dried particles were subject to visual and quantitative analysis using nano-focused x-ray tomography to allow a three-dimensional structural description. Extracted morphological descriptors of each particle have been related to solid phase nucleation and growth kinetics. Special emphasis of this study is the evaluation of the solid phase porosity, structural thickness and crystallinity influenced by the solidification from solution and with an impact on key particle performance attributes such as mechanical stability, manufacturability, solubility and solid state stability. Formulations of an active pharmaceutical ingredient with polymeric excipients have been developed to address challenges of a highly crystalline solid with large elongated primary particles. The formulated particles exhibited significantly improved structural modifications leading to an increased mechanical stability, (partially) suppressed crystallinity and tailored dissolution rates. The overall observed diversity of particle properties obtained from single droplet experiments for changing compound systems expressed the importance of an in-depth characterisation of the solidification process prior to process implementation.

In summary this study combined the capabilities of an acoustic levitation system and a nano-focused x-ray tomograph to provide a platform to investigate the particle formation and allows a morphological characterisation on a single droplet scale. A comprehensive study enables the development of a mechanistic understanding for a rational particle design approach of novel compound systems targeting critical particle properties influencing particle performance. The fundamental knowledge gained from this method can then be transferred to develop a design space for process applications involving particle formation and drying such as spray drying.

508011 Effect of Casting Techniques on Critical Quality Attributes of Strip-Films

Abstract:

This work is part of an overall objective of being able to create patient compliant drug loaded thin polymeric films that are uniform in drug content or where the amount of drug varies by a minuscule amount and has negligible effects to the drug dosage. The goal of this specific work is to compare the critical material attributes (CQAs) of strip-films, loaded with a BCS Class II drug, produced via solution casting and slurry casting. As-received fenofibrate (FNB) was used to prepare slurry casted films as the baseline/control for CQAs. The primary difference between the two techniques is that the drug is fully dissolved in solution casting, as opposed to slurry casting where the drug is suspended in the polymer solution as nanoparticles or micronized coated particles. Solution casting may increase the dissolution rate of the poorly water-soluble drug producing amorphous particles while in the slurry casting introducing nano-sized or micon-sized drug particles into water based system increases the dissolution rate with highly increased total surface area of the drug. For both methods, FNB was used as the model poorly water-soluble drug, hydroxypropyl methylcellulose (HPMC) E-15 as the film-former, and glycerin was used as the plasticizer. For solution casting, a binary solvent system consisting of 1:4 ratio of deionized water to organic solvent was used. Three organic solvents were evaluated: methanol, ethanol, and acetone. FNB nanosuspensions were prepared via wet stirred media milling (WSMM). Micronized coated FNB particles were prepared via fluid energy milling (FEM). Mechanical attributes (Youngâ€™s Modulus, tensile strength, and elongation at break), dissolution rate of the drug, content uniformity, and crystallinity of the drug in each film were evaluated as the CQAs of the strip-films.

All the films exhibited similar dissolution profiles of films loaded with nanosuspension and solution films. While as was expected, the films loaded with micronized coated powder released slower. Overall, the film CQAs containing the drug nanosuspension and micronized coated powder were found to be superior. Those films had better drug content uniformity, higher elongation at break, and long-term stability; all of these are desirable CQAs. Interestingly, solution casted films had acceptable drug content uniformity, despite re-crystallization of the drug, and visual non-uniformities. Unfortunately, solution casting had a host of manufacturing problems such as longer drying time due to need for lower maximum drying temperatures in order to avoid excessive bubble and defect formation, unpredictable drug particle re-crystallization and different drug particle shapes that change with time, and most importantly, burden of formulation development that also includes selection of solvent that must be tailored to a specific drug.

507853 Experimental Studies on the Effect of Moisture Content and Volume Resistivity on Electrostatic Behaviour of Pharmaceutical Powders

Abstract:

Pharmaceutical powders are mainly organic materials and are likely to be charged due to repeated inter-particle and particle-wall contacts during industrial processes. This study experimentally investigated the effect of moisture content (ranging from approximately 1.8 to 30 wt. %) on tribocharging behaviour of pharmaceutical granules, as well as their apparent volume resistivity. The tribocharging behaviour of pharmaceutical granules was investigated using a rotating device and apparent volume resistivity was measured in a conventional volume resistivity test cell. Additional measurements were performed on individual ingredients, each having the same moisture content as that of the granules, in order to investigate the effect of each single ingredient on the apparent volume resistivity of granules. In this work, the individual ingredients used for granules were: -Lactose Monohydrate (-LMH), Microcrystalline Cellulose (MCC), Hydroxypropyl Methylcellulose (HPMC), and Croscarmellose Sodium (CCS). The results showed that the specific charge of granules began to increase at the moisture contents below 5 wt. %, which can be referred as critical moisture content of granules. The apparent volume resistivity showed the same behaviour, indicating that the specific charge could be due to an increase in apparent volume resistivity of granules at reduced moisture content. Finally, it was shown that the apparent volume resistivity measured for granules was mainly affected by that of the -LMH, the major component of granules accounting for 40 wt.%.

508071 Formation of Gradient Particle Pattern Via Electric Field Guided Electrospray Deposition

Abstract:

Electrospray and electric-hydrodynamic atomization have been investigated in recent years for the production of pharmaceutical particles. Furthermore, there were literature report on the electrospray deposition for fabrication of various patterns based on different types of materials such as living cells, proteins, and pharmaceutical particles. In this project, the effect of the configuration of grounded electrode used in electrospray was investigated. Three types of grounded electrode, namely, point, line and plane electrode were placed under a glass slide. The electric field and particle deposition were simulated by solving the Poissonâ€™s equation and the force balance equation for the charged particle during the electrospraying process. It was found that the particle patterns with graded size can be obtained when the additional point or line electrode was used. The formation of gradient patterns was ascribed to the gradient of electric field and the corresponding focusing effect. By changing the electric field, the feature size of the gradient patterns can be controlled. In addition, an analysis of the particle patterns obtained by using different collecting substrates, e.g. normal glass slide, Polydimethylsiloxane (PDMS) coated glass slide, showed that the particle pattern size is much more uniform on PDMS â €“coated glass slide under the same condition. This kind of gradient patterns may be promising for applications in high throughput drug screening or biomaterial screening for studying cell material interactions.

506444 Relative Importance of Cohesion and Internal Friction in Flowability of Pharmaceutical Powder

Abstract:

The flow function, measured by shear cell testers, has become the standard metrics for assessing powder flowability in the development and manufacturing of pharmaceutical solid dosage forms. Commercial shear cell testers calculate flow function by determining a linear yield locus through measurements of yield strengths of materials under different normal stresses. Cohesion and internal friction are two quantities defined in Mohr-Coulomb theory that describe the linear yield locus; hence they also determine flow function. The purpose of our study is to understand the relative contribution of cohesion and internal friction to flow function and to apply the findings to improve powder flowability during formulation development.

Through the analysis of 1130 ring shear tests of various pharmaceutical powders, we found that the flow function is strongly correlated with the ratio of pre-consolidation stress to cohesion. The powders included single-component or multi-component system of excipients and APIs that were prepared by simple blending or granulation. However, flow function had almost no correlation with internal friction angles.

We developed a model based on the Mohr-Coulomb theory to verify the experimental findings. The above observationâ €"flow function strongly correlates with the ratio of pre-consolidation stress to cohesion but not with internal frictionâ €"was a mathematical consequence ensued from Mohr-Coulomb failure analysis. The model suggested that if the powders exhibited a small difference between the angle of internal friction at steady-state flow and the angle of internal friction at incipient flow, the correlation remained valid. Most pharmaceutical powders, coincidentally, possess such characteristic. Therefore, in practice, cohesion has a strong correlation with flow function while internal friction has none.

We demonstrated an application by mixing colloidal silicon dioxide with Hypromellose powder to show that substantial increase in flow function was accompanied by large decrease of cohesion while the internal friction angle showed negligible change. In contrast, mixing colloidal silicon dioxide with anhydrous calcium phosphate powder has no effect in flow function when cohesion changed minimally. Practitioners in solid dosage drug development can apply this knowledge and focus on improving powder flowability by reducing powder cohesion instead of internal friction.

[1] This study has been published in Leung et al. 2017. *Journal of Pharmaceutical Sciences*. 106:1865-1873

507030 Evaluation of Mechanical Properties of Organic Crystals for the Prediction of Breakage during Isolation Processes

Abstract:

Crystal breakage during the isolation of active pharmaceutical ingredients (API) is an issue of great concern to the pharmaceutical industry. Conservation of the desired particle size distribution (PSD) throughout downstream processing is extremely important as changes are known to affect properties such as bioavailability, flowability and bulk density [1]. APIs are commonly produced by batch crystallisation as high aspect ratio crystals, usually needle-shaped, and they are separated from the mother liquor through pressure filtration. At industrial scale, isolation of the material is prone to significant changes in the PSD. This is typically attributed to agitation during drying [2]. In contrast, MacLeod and Muller [3] showed that the filtration process itself can have a profound impact on the PSD of needle-shaped crystals. The main objective of this work is to elucidate the phenomenon of crystal fracture under pressure filtration. To achieve this, mechanical properties, such as yield stress, elasticity and hardness, need to be determined at the single crystal level.

Namazu et al. [4] conducted 3-point bending test experiments using an inorganic single crystal (Si) fixed beam in order to evaluate the size effect on the mechanical properties of crystals. It was found that reduction of the crystal size leads to an increase in bending strength. Our group developed a single crystal 2-point bending test by fixing one edge of single microscopic needle-shaped crystals and applying a force at the other edge using Atomic Force Microscopy [5]. We employed this method to determine the micro-mechanical properties of β-L-glutamic acid (β-LGA), a needle-shaped organic crystal. Similar to [4], a high number of experiments demonstrated that the yield stress is a strong function of the crystal size. Using a derivative from Bernoulli's beam theory [3], it was found that the critical stress leading to β-LGA crystal snapping is equal to 13.8 MPa. As the crystals are by nature not identical, a Weibull analysis was also performed to model the distribution of the experimental observations. The Weibull's distribution function shape (k) and scale (λ) parameters were estimated and are equal to 2.07 and $1.896\text{x}10^7$ respectively. The median of the Weibull distribution is equal to 15.9 MPa, meaning that 50% of the crystals would break under such a stress [6].

A solid mechanics model, implemented in COMSOL Multiphysics, was used to simulate the single crystal fixed beam bending. The model can very accurately predict the crystal deflection and stress profiles under the application of varying forces. The extension of such a model to carry out simulations at the filter bed scale, in conjunction with yield stress experimental findings, could be ultimately used as a predictive tool towards the evaluation of the propensity for particle breakage during API isolation at scale, and the improvement of the process design as well as operating conditions.

References

1. Kougoulos, E., C.E. Chadwick, and M.D. Ticehurst, *Impact of agitated drying on the powder properties of an active pharmaceutical ingredient.* Powder Technology, 2011. **210**(3): p. 308-314.

2. Lekhal, A., K.P. Girard, M.A. Brown, S. Kiang, J.G. Khinast, and B.J. Glasser, *The effect of agitated drying on the morphology of L-threonine (needle-like) crystals.* International Journal of Pharmaceutics, 2004. **270**(1-2): p. 263-277.

3. MacLeod, C.S. and F.L. Muller, *On the fracture of pharmaceutical needle-shaped crystals during pressure filtration: Case studies and mechanistic understanding.* Organic Process Research and Development, 2012. **16**(3): p. 425-434.

4. Namazu, T., Y. Isono, and T. Tanaka, *Evaluation of size effect on mechanical properties of single crystal silicon by nanoscale bending test using AFM.* Journal of Microelectromechanical Systems, 2000. **9**(4): p. 450-459.

5. Sohi, S.S., S.D. Conell, D. Harbottle, and F.L. Muller, *A method for the determination of mechanical properties of needle shaped crystals using Atomic Force Microscopy.* Master's Report, University of Leeds, 2016.

6. Hallac, F.S., I.S. Fragkopoulos, S.D. Connell, and F.L. Muller, *Micro-mechanical properties of needle-shaped organic crystals.* Organic Process Research and Development, 2017. *To be submitted.*

CHARACTERIZATION OF PARACETAMOL GRANULES FORMED BY BINDER DROPPING

Dr. Sheena M. Reeves, Prairie View A&M University, Prairie View, TX

Abstract

Particulate materials are widely use in pharmaceutical industry in drug formulations. Some formulations are made using the process of agglomeration or granulation. Most active pharmaceutical ingredients (APIs) are administered in a formulation containing excipients or binders to prevent segregation. Excipients are inert materials that are used to assist drug delivery; however, these additives sometimes cause mild adverse effects and hinder API dissolution. Glycerides and waxes are added to most APIs to increase stability. Drug bioavailability and solubility is a constant challenge with pharmaceuticals. A variety of excipients exist and serve various purposes including added stability, segregation prevention, and controlled dissolution. Binding agents under consideration are classified as semi-synthetic or synthetic and include a comparison of polyvinylpyrrolidone (PVP) and polyethylene glycol (PEG). Although it is considered biologically inert and safe, PEG is known to cause allergic reactions to some individuals. Since PEG is a common binder, a replacement binder should have similar characteristics and capabilities. The purpose of this research is to investigate 1) how binding agents influence the ability of a paracetamol granule to dissolve in a buffer solution (pH 7.0) and 2) the ability of the granule to maintain cohesiveness during compression testing.

Wet granulation was chosen as the method of granule formulation and was accomplished by dropping solutions onto a powder bed. Initial tests in this project involved developing paracetamol granules of similar size, shape, and weight for consistency and uniformity. Upon drying, images of the granules were analyzed using Image-Pro Premier software and included measurements for perimeter, aspect ratio, area, diameter, and roundness. To create a granule containing between 275 – 300 mg, five drops of binder were used. To test for granule stability, compression tests were conducted using a conical indenter and a flat surface. Granules produced from synthetic binder solutions proved more cohesive and stable. Five binder drops produced granules without excess agglomeration around the edges with an average weight around 300 mg. Dissolution rates were determined using an Agilent Bio-Dis Apparatus 3 and UV-Vis spectrometry. The concentration of the binder, the type of binder, and the dip speed of the apparatus were varied. Pure paracetamol powder or granules formed with deionized water, a very weak binder, served as a baseline for analysis. Further investigation of the granule cohesiveness was conducted using scanning electron microscopy (SEM).

507694 Investigation of Single-Droplet Drying Kinetics in an Acoustic Levitator, Micro-Sphere Morphology and Mechanical Properties of Pharmaceutical Excipients

Abstract:

In the last years there has been significant interest in advanced continuous manufacturing of pharmaceuticals. These manufacturing routes have significant advantages, including a smaller plant footprint (smaller equipment and facilities), reduced manufacturing time, elimination of unnecessary storage by streamlined production from start to finish, and lower manufacturing costs. Moreover, quality is ensured in real time, leading to a better overall efficiency of the manufacturing process.

One of the challenges in continuous production is the link between primary (API synthesis, crystallization, purification, filtration and drying) and secondary manufacturing (blending, granulation and tableting/capsule filling). Particle properties of crystalline APIs are tailored in primary manufacturing by wet chemical processes. To enable further processing of the crystalline API in secondary manufacturing, a drying step is necessary to remove the residual solvents.

One route to transform a fluid feed stream into dry powders is spray drying. A major benefit of spray drying over other technologies is that drying is conducted close to the wet bulb temperature of the slurry, and therefore it is suitable for heat-sensitive products. As long as the surface of a droplet/particle is covered with a liquid layer, drying occurs at the wet bulb temperature in the first drying stage. This liquid layer is removed by evaporation of the liquid phase of the slurry into the drying air until solidification occurs. Afterwards, drying progresses in the second drying stage at temperatures converging to the dry bulb temperature until the designed residual moisture content is reached.

In suspensions, solids are dissolved in the carrier liquid and suspended as particles. The drying of suspension slurries results in precipitation of dissolved material forming bridges between the primary particles. Therefore, the final microsphere properties are influenced by the amount of dissolved solids, since this determines the size and amount of bridges between primary particles. There is a clear correlation between more dissolved material in the slurry liquid and the formation of stronger bridges between primary particles. This alters significantly the final microsphere properties, as well as the corresponding processability.

In this work we focus on the correlation between slurry composition and the resulting material properties after drying. Drying experiments were conducted by acoustic levitation of a model compound in binary liquid mixture droplets. The acoustic levitation setup enabled microscopic observations of the drying behavior of single droplets. Variations in the suspension composition revealed the impact on the drying kinetics as well as the mechanical properties and the morphology of the final particles (microspheres). Different microsphere morphologies were observed, from very loose agglomerates to densely packed microspheres with crust formation. Electron micrographs of the microsphere morphology of different suspension compositions were in good agreement with the hardness measured by a compression strength test.

508392 Enabling Direct Compaction at High Drug Loading Via Dry Coating of APIs: Towards a Predictive Framework

Abstract:

High speed direct compression (DC) tabletting is widely considered as the most ideal process to produce tablets in the pharmaceutical industry. DC continuous processing route is a natural and economical link between continuous API and drug manufacturing. It is beneficial because it consumes lower energy, resources, time and money compared to alternative processes that may include extra laborious unit operations for granulation and drying. However, due to the cohesive nature of active pharmaceutical ingredients (APIs), achieving drug loading of 30 % or higher via DC route is not possible in batch or continuous processing. These problems get more severe when APIs are micronized for enhanced content uniformity (CU) or bioavailability improvement, and the use of specialized higher performance excipients do not necessarily help. Dry coating of cohesive APIs via solventless continuous process has been proven to significantly improve the flowability of cohesive micronized APIs, as well as the resultant blends. Dry coating with silica is also shown to minimise adverse impact of the native surface properties, e.g., surface energy and roughness, of APIs, leading to a more robust toolbox for property enhancement. Most recent NJIT work targets dry coating for enabling DC at higher drug loading. Our results show that dry coating significantly improves the pharmaceutical blend properties and allows higher API loading into blends. Furthermore, the feasibility of using finer excipients, which helps in reducing segregation tendency and enables better compaction, to formulate high drug loaded blends with excellent bulk properties has been demonstrated. Blends formulated with dry coated micronized APIs and finer excipients consistently had better flowability compared to the same APIs blended with coarse excipients, which contradicts current industrial practices. On the other hand, for very low drug loading, drug content uniformity greatly improved with dry coated micronized APIs. In order to make wider use of DC route as a link between continuous API and drug manufacturing, it is necessary to develop a modelling framework is necessary to predict API and blend properties such as packing and flowability. NJIT team is developing such framework that can allow the determination of the highest API loading for a given set of properties of the API manufactured via continuous manufacturing with and without API dry coating. Alternately, given the target API loading, such models can determine the API size that must be produced during CM, if they need to be dry coated, and if special purpose excipients would be required.

508641 Particle Engineering at the Drug Substance, Drug Product Interface â€“ Inherently Continuous Processes Yielding Advantageous Material Properties

Abstract:

Results will be presented on a three tiered approach to achieve advanced drug product intermediates addressing the inherently challenging properties of poorly soluble APIâ€™s. These include:

1) the incorporation of excipients during particle formation and subsequent isolation and drying of drug substance processing, which has historically been a vastly underrepresented area of development in the pharmaceutical industry

2) the use of a novel, high shear mixing device to allow a revised approach to precipitation to achieve either nano crystalline materials or amorphous solid dispersions

3) subsequent particle isolation utilizing a thin film evaporator, which yields additional capability to modify individual particle (i.e. surface characteristics and local dissolution rates) and bulk level (powder flow and compressbility) properties through selection of process parameters and/or further incorporation of excipients.

Ultimately this allows for the design of complex materials using a continuous, volumetrically productive process. Additionally, the drug product intermediates afford streamlined drug product processing, including continuous direct compression. Work will be presented for multiple component systems achieving high drug load intermediates comprised of 1) co-precipitated amorphous solid dispersions (cPAD) and 2) nano crystalline API.

508056 The Â Role of Surface Energy on Developing Engineered Excipients during Dry Coating Process

Abstract:

Abstract

Excipients with good flowability, bulk density as well as compaction properties are desired for use in tableting since they play important roles in formulation development and processing, including, handling, mixing, feeding and compaction. The objective of this work is to examine the feasibility of using dry coating based surface modification of microcrystalline cellulose, Avicel PH-105, to produce an engineered fine grade (particle size < 30Î¼m) excipient that has all three desired properties. Using a material sparing, model based dry coating methodology, as-received grades of MCC were dry coated with 1wt% silica R972P, Aerosil 200 and M-5P. The results indicate that as expected, the bulk density and flowability of dry coated MCCs were significantly improved, while there was an appreciable loss of compaction. To understand the loss of compaction, the surface energy of the three grades silica and dry coated products were measured by using inverse gas chromatography. The results show that the loss in compaction is related to the reduced dispersive surface energy of the dry coated powders. Remarkably, use of Aerosil 200 led to a minimal reduction in compaction among these three grades of silica due to its highest dispersive surface energy among the selected silicas. This study shows that a new high functional grade of fine excipient which has a good flowability, bulk density and excellent compaction property was successfully developed. Moreover, surface energy can be a useful tool to predict whether a material can be good candidate for related studies in future.

507505 A Mechanistic Approach to Developing Spherical Agglomeration for Process Intensification in Pharmaceutical Manufacturing

Abstract:

The crystallisation of particulate materials is a common process with applications in a variety of industries. This is often the first operation to recover crystalline material from a bulk solution, thereby presenting an opportunity to beneficially tailor material properties. In the pharmaceutical industry, the process is already used to facilitate the formulation of the majority of active pharmaceutical ingredients (APIs). Currently, the formulation and structure of the APIs directly dictate the required number of unit operations to follow. For solid oral dosages, it is imperative that the API has high flowability and compressibility to ensure an appropriate tablet structure is produced. Furthermore, these properties are essential for predictable API performance within the body. To improve these properties, the industry relies upon granulation and milling operations which are both time-consuming, cost-intensive and energy-inefficient. They are also difficult processes to accurately control, which may involve the recycling of crystalline material which is unsuitable for tabletting.

Spherical agglomeration is a process which provides the opportunity for these key properties of APIs to be significantly improved through the direct precipitation and agglomeration of crystalline structures. The precipitation is achieved by addition of a poor solvent (anti-solvent) to a good solvent in which the drug is dissolved and the agglomeration is supported by addition of bringing liquid. These steps can follow on from each other or be achieved simultaneously, often with improved results [Wu et al., 2015]. The particle size can be increased by a factor of ten, compared to the original size [Kawashima et al., 1982]. Additionally, the agglomerates formed are often dense and spherical in nature; a major improvement for naturally acicular particles. As a result, if the process is optimised, there is the potential for several unit operations to be removed from the production process. The process could therefore significantly intensify the manufacture of pharmaceutical products through the reduction of the time to market, reduction of the operating capital required and reduction of equipment requirements. There is also the potential for the system to be operating continuously, further improving API production efficiency.

The current state of the art has already established robust relationships between processing parameters and agglomerate properties. These are consistent, regardless of the crystalline product to be agglomerated. This provides engineers with greater flexibility in identifying a set of process parameters which will yield agglomerates with the required size, porosity and strength, amongst other characteristics. However, for this flexibility to be utilised effectively, it is vital that engineers have reliable models which can accurately predict process performance. Currently, such models do not exist due to a lack of key mechanistic understanding. Whilst previous studies have tried to identify and model some of these mechanisms, no one study has utilised all the mechanisms to produce a model which accurately reflects the process.

Critically, three separate overarching rate processes occur; wetting, growth and breakage. The wetting phase can occur through either the distribution or immersion mechanism, depending on the size of the particles to be agglomerated and the size of the droplets of bridging liquid which agglomerate them. If the droplets are smaller than particles, a distribution mechanism occurs in which the particles are effectively wetted through a coating mechanism, allowing particles to stick together. If the droplets are larger than particles, an immersion mechanism occurs in which the particles move inside the droplets and stick together within it. The latter of these mechanisms gives rise to denser, more spherical agglomerates [Muller & Loffler, 1996]. The growth phase involves the compaction and consolidation of agglomerates as they are subjected to shear forces. Crucially, as more particles adhere to each other, agglomerates growth in size through the creation of liquid bridges. These solidify to yield crystalline bridges upon drying. The breakage phase is not investigated in this study, but occurs as particles collide with each other and the crystallisation vessel.

In this study, we report our initial work in identifying, characterising and modelling of wetting/nucleation and growth/consolidation mechanisms. To do so, the fundamental kinetics were identified and verified using model particles; plastic beads (52 μm) and glass ballotini beads (45-90 μm). A novel, microfluidic device was developed to visualise these mechanisms in-situ with an optical microscope and a high-speed camera. We have also investigated the influence of material properties, such as surface energy, and process parameters, such as shear rate, upon these mechanisms. From these experiments, will have been able to construct the beginnings of a rate process map, which demonstrates the relationship between these parameters and their effect upon the process as a whole.

The roles of such mechanisms and phases of agglomeration using crystal loaded systems, such as salicylic acid, are also presented. Here, two unique pieces of equipment are employed; an oscillatory baffled reactor (OBR) and a particle viewing and measurement (PVM) probe. The OBR has superior mixing capabilities compared to conventional stirred-tank reactors due to the elimination of dead zones. The level of mixing can be analysed computationally through fluid dynamics. The PVM allows the analysis and tracking of particle populations in real-time, allowing us to further understand the agglomeration process with a view to improving and developing prior determined models.

S. Wu, K. Li, T. Zhang, J. Gong, Size Control of Atorvastatin Calcium Particles Based on Spherical Agglomeration, Chem. Eng. Technol. 38 (2015) 1081–1087.

Y. Kawashima, M. Okumura, H. Takenaka, Spherical Crystallization : Direct Spherical Agglomeration of Salicylic Acid Crystals During Crystallization, Science. 216 (1982) 1127–1128.

M. Müller, F. Löffler, Development of agglomerate size and structure during spherical agglomeration in suspension, Part. Part. Syst. Charact. 13 (1996) 322–326.

507700 Morphological Stability of Microencapsulated Lipophilic CompoundsÂ By AFM Imaging

Abstract:

Micronutrients and nutraceuticals such as vitamins, carotenoids and polyunsaturated fatty acids (PUFAs) are essential for human and animal health and well-being. Fat-soluble vitamins, like Vitamin A, or PUFAs are often chemically instable (*e.g.* to oxidation, heat) and they cannot easily be added to a hydrophilic food environment like a beverage. Thus, such vitamins are mostly sold in microencapsulated product formulations to protect the active ingredient from the surrounding environment. The morphological stability of microencapsulated lipophilic compounds directly influences the long-term storage stability and performance in final applications. Thus, during the development of novel product formulations, the stability of the encapsulated active ingredient is typically monitored in time-consuming studies, storing the product formulations at different environmental conditions and assessing the concentration of the active ingredient at several time points. Here, a facile method to investigate the internal morphology and structural stability of microencapsulated vitamins has been successfully developed. The internal particle morphology of three commercial dry Vitamin A acetate formulations differing in their matrix composition (hydrocolloid/polysaccharide and plasticizer) was systematically characterized by atomic force microscopy (AFM). The evolution of the particle internal morphology after *per-se* storage at 25 and 40 °C for ten months was investigated. Furthermore, an accelerated stress test by *in-situ* heating and AFM imaging was developed. Surface roughness parameters from the AFM images were used to semi-quantitatively estimate the evolution of the particle internal microstructure, and thus the aging and destabilization of the lipophilic vitamin formulations. The influence of humidity level on the thermal morphological restructuring of one of the commercial samples was also investigated. This study has shown that such AFM investigations of microparticles are a novel analytical tool to rapidly estimate the physical stability of the lipophilic vitamin dried emulsions.

Mougin, K., A. Bruntz, D. Severin and A. Teleki, "Morphological stability of microencapsulated vitamin formulations by AFM imaging," Food Structure 9, 1-12 (2016).

508208 Multi-Sensor Measurements of Quantitative Particle Size and Shape Information in Crystal Slurries

Abstract:

Introduction

Particle size and shape are critical quality attributes for active pharmaceutical ingredients (API) as they have direct impact on downstream processing, as well as on the performance behaviour of the finished product. Obtaining a reliable and robust quantitative information of these particle attributes in real-time remains a great challenge across the multiple manufacturing steps and cannot be achieved by just using a single sensor.

In-line particle sizing technologies, such as Focus Beam Reflectance Measurement (FBRM) and Particle Vision and Measurement (PVM), measure the optical response of single particle (single scattering) to a single wavelength light source and measurable size ranges are limited by the optical resolution of the light source and the optics. The increase of turbidity in the medium leads to an increase of scattering effects and absorption contributions. When photons encounter several particles, they undergo multiple scattering, changing the direction during each scattering event. In these cases, it is very challenging to obtain reliable results from either FBRM or PVM. Spatially and angularly-resolved diffuse reflectance measurement (SAR-DRM) is a spectroscopic-based approach for in-line monitoring.[1] SAR-DRM collects multi-wavelength (UV-visible-NIR) diffuse reflectance spectra from optical fibres of multi-angle multi-space arrangements. Absorption of light occurs at specific wavelengths for different particles, which can be used to identify samples composition; and different particle sizes and shape will have a different scattering pattern. SAR-DRM spectrum from each angular and spatial arrangement yields differences responses to the changes in the scattering and absorption properties of the sample. The technology requires multiple scattering on the incident light, therefore, it could complement other single particle based measurement technologies for in-line process monitoring.

In this work, crystal slurries are quantitatively characterised by using an in-line multi-sensor approach. Multivariate regression analysis is employed to establish calibration models for estimating particle size and solid loading. The knowledge acquired is then applied to a wet milling process, which is commonly used in pharmaceutical industries to improve the solubility and bioavailability of poorly water-soluble drugs by causing the breakdown of coarse particles into finer ones through mechanical forces.

Methodology

1. Characterisation of particle attributes, evaluation of the optical probes and modelling strategies

The investigation was carried out on crystal slurries of metformin HCL of different particle size ranges (<53, 53-90, 90-125, 150-180, 180-250 and 250-355 μm) and solid loadings (from 1 up to 40 wt.%). The samples were monitored using FBRM, PVM and SAR-DRM. Mathematical algorithms were applied to FBRM and PVM data to extract particle size distribution (PSD) and aspect ratio [2, 3]. Prior to each experiment, samples were analysed using off-line technologies such as laser diffraction and imaging.

Characterisation of the particles attributes served as an input to validate SAR-DRM sensitivity, accuracy and capability to track the differences in size and solid loading in crystal slurries. Multivariate analysis was applied to establish a performance matrix for all probes and combined methods in order to acquire a reliable and robust model to estimate particle size and concentration.

2. Application of the strategy to a wet milling process

A wet mill was coupled to the 250ml vessel via a recycle-loop configuration in gently suspended conditions in order to perform measurement and extract samples for off-line measurements. Breakage of metformin crystals through this wet milling method produced a range of end sizes from different starting particle sizes and concentrations. The significant change in particle attributes undergoing this system was then continuously monitored by FBRM, PVM and SAR-DRM. The same multivariate analysis was applied to assess the robustness of the model to this complex system.

Conclusions

The results show that FBRM and PVM are valuable tools to monitor particle size and shape distributions in low dense mediums. Increasing the solid loading, regardless of the particle size used, particles start overlapping and pictures lose their resolution which leads to ambiguous results in both FBRM and PVM. Larger particles are more likely to be overlapping or touching the picture frame and consequently, more likely will be discarded by the PVM imaging algorithm. In case of FBRM, large particles generate chords split, shifting the CLD towards smaller lengths.

SAR-DRM is capable of handling high solid loadings and relatively large particle sizes and is sensitivity to both particle size and solid concentrations, performing better than FBRM and PVM for such suspensions. SAR-DRM can be a complementary technique to the current in-line particle analysis to achieve robust process monitoring and enable improved control & optimisation of crystallisation processes.

References:

[1] Thennadil. Suresh and Yi-Chieh. Chen, Measurement apparatus and method, U.S. Patent EP2783199A1 (2014).

*[2] Agimelen. Okpeafoh, Hamilton. Peter, Haley. Ian, Nordon. Alison, Vasile. Massimiliano, Sefcik. Jan, Mulholland. Anthony, Estimation of particle size distribution and aspect ratio of non-spherical particles from chord length distribution, Chemical Engineering Science, **123**, 629-640 (2015).*

*[3] Agimelen. Okpeafoh, Jawor-Baczynska. Anna, McGinty. John, Dziewierz. Jerzy, Tachtatzis. Christos, Cleary. Alison, Haley. Ian, Michie. Craig, Andonovic. Ivan, Sefcik. Jan, Mulholland. Anthony, Integration of in situ imaging and chord length distribution measurements for estimation of particle size and shape, Chemical Engineering Science, **144**, 87-100 (2016).*

508226 Monitoring API Crystal Breakage in Wet Milling Using Inline Imaging and Chord Length Distribution Measurements

Abstract:

Introduction

Particle size and shape are critical attributes that affect the performance of particulate products. The monitoring and control of particle size and shape is an ongoing challenge in the field of particle technology. Current practices for determining particle size and shape mostly rely on offline analysis, in which the measurement is carried out after the process has finished. In addition to the offline analysis being carried out at the end of the process, the particle size and shape may be altered by the filtration and drying process required to extract particles from a liquid medium. Hence, the particle size and shape determined at the end of the process may not be representative of the size and shape of the particles during the process.

The Particle Size Distribution (PSD) and shape (quantified by the aspect ratio - ratio of particle width to length) of the particles in a suspension can be monitored in situ using data from suitable instruments. The Focus Beam Reflectance Measurement (FBRM) and Particle Vision and Measurement (PVM) sensors are inline probes that can be used for this purpose. The FBRM sensor measures a chord length distribution (CLD) which depends on the PSD and aspect ratio of suspended particles. The PSD and aspect ratio of the particles can subsequently be estimated from the CLD data by solving an inverse problem [1]. The PVM sensor captures images of particles in a suspension from which the PSD and aspect ratio of the particles can be estimated using a suitable image processing algorithm [2].

In this work, we employ previously developed algorithms [1,2] for solving the inverse problem associated with CLD data and analyse images from the PVM sensor to monitor the PSD and aspect ratio of particles during wet milling of organic crystals. Wet milling is an emerging application carried out in industry to reduce the size of particles to a suitable range. Organic crystals typically undergo breakage from wet milling. However, various systems and process conditions produce different effects which pose different challenges to the CLD inversion as well as the image analysis. We study the strengths and limitations of the CLD inversion and image processing under various conditions and find that they complement each other in obtaining quantitative information about particle size and shape.

Materials and Methods

A 1L stirred tank vessel charged with different particle suspensions was connected to an IKA MagicLab Wet Mill (a rotor-stator mill) in a recycle-loop configuration. The stirred tank was equipped with temperature controller and the FBRM and PVM probes were used for data capture. The suspension was passed through the wet mill at a series of increasing rotor speeds with real time monitoring. Benzoic acid, paracetamol and metformin hydrochloride were used in three separate experiments. In each process, the CLD data at every stage was captured with the FBRM probe while images of the particles were acquired using the PVM probe. Before the start of each process, samples of each material were analysed using the Malvern Morphologi G3 instrument (offline particle imaging) for particle size and aspect ratio estimation. Similarly, samples of the filtered and dried milled particles were analysed for particle size and aspect ratio estimation at the end of each wet milling process using the Morphologi G3 instrument.

Results

Obtained results from the employed method has shown the relative sensitivities of the two modalities to changes in particle size and shape due to the wet milling. The effect of bubbles produced during the process on the two modalities has also been studied. When the particles are sufficiently large and within the resolution of the PVM instrument camera, and not too large so as to into the image frame, the inline imaging method gives PSD estimates which are closer to online estimates of similar materials. This is especially true for systems composed of long needle-like particles such as benzoic acid and metformin HCL mixed with shorter or more rounded particles. This is because the CLD method needs to find a compromise PSD which the measured CLD at an appropriate aspect ratio. In systems like these, the aspect ratio distribution estimated by the imaging method tends to be dominated by the smaller particles, while the mean aspect ratio estimated by the CLD method tends to be dominated by the long needle-like or rod-like particles. However, the inline imaging method is limited to particles of sizes above about 30Î¼m, whereas the CLD method can go to smaller sizes. In addition, the PSD estimated by the inline imaging method becomes less representative as the sizes of the particles approach the size of the image frame. The accuracy of the PSD estimate is also affected by the proportion of objects that are captured outside the focal plane of the camera. Similarly, the PSD by the CLD method becomes less representative as the length of needle-like or rod-like particles approach the diameter of the circular trajectory of the FBRM laser spot making the chord length probability estimate 22 less accurate. Hence, in systems composed of a mixture of needle-like or rod-like particles and more rounded particles of various sizes, a combination of both the CLD and inline imaging methods (probably in a multi-objective approach) should give more robust estimates of the PSD and the aspect ratio. This will be particularly important to real-time monitoring and control of crystallisation processes where various process conditions could lead to the production of particles of various sizes and shapes and effects such as bubbles which could be very challenging to capture by a single sensor method.

Conclusions

We have employed wet milling processes of slurries of different crystalline materials to assess the strengths and limitations of two different inline particle monitoring modalities namely CLD and PVM imaging. The materials were carefully chosen as they produce crystals of different morphologies and mechanical strengths. The results from both CLD inversion and PVM image analysis demonstrate the ability of our software tools to quantitatively monitor particulate processes with respect to particle size and shape distributions. The results from the three case studies have identified a number of strengths and limitations of each method depending on the process conditions and nature of particles in each system. The CLD inversion algorithm can be applied to dense particle suspensions whose images may be difficult to analyse. However, the CLD data is affected by effects such as bubbles or particle transparency. The image processing algorithm allows the distribution of aspect ratios of particles to be estimated, but it is limited by the resolution of inline images

References:

*[1] O. S. Agimelen et al., Estimation of particle size distribution and aspect ratio of non-spherical particles from chord length distribution, Chemical Engineering Science **123**, 629 (2015).*

*[2] O. S. Agimelen et al., Integration of in situ imaging and chord length distribution measurements for estimation of particle size and shape, Chemical Engineering Science **144**, 87 (2016).*

508703 Investigating Surface Energy Heterogeneity of Processed Lactose Powders

Abstract:

Particle surface properties are known to be affected by the processing routes and processing conditions used for generating particles. Inverse Gas Chromatography (IGC), particularly Finite-Dilution Inverse Gas Chromatography (FD-IGC), has been shown to be a versatile technique for surface energy characterisation of solids. In this study, lactose powders processed by both milling and sieving the similar crystalline Î±-lactose monohydrate starting material were used. The sieved samples used in this study were divided into top, middle and bottom fractions. FD-IGC was used to examine the surface energy heterogeneity of these processed crystalline Î±-lactose monohydrate powders. The dispersive component of surface energy at different surface coverages of lactose was determined in this work. The results indicated that all lactose samples investigated showed some degree of heterogeneity. Unmilled, coarse lactose powders showed lower dispersive surface energy values and fine powders, produced by milling, were found to have the highest dispersive surface energy values. The surface energy distribution obtained from the deconvolution of the surface energy data obtained from FD-IGC using a Boltzmann distribution model and simulation-fitting approach suggested possibly three different surface energy regions on the processed lactose samples. The surface energy region with highest dispersive surface energy ranges from 44-49 mJ/m^2 followed by a 38-45 mJ/m^2 region and a region with lowest dispersive surface energy of 26-31 mJ/m^2. For sieved samples, the surface energy distributions for the top and medium sieved fractions suggested that the surface energy of large, unbroken Î±-lactose monohydrate crystals consists of roughly equal amounts of higher and lower energy surfaces whereas, the bottom fraction with smaller particles had greater proportion of high energy sites and less of the lower energy sites than the other two sieved fractions. Overall, FD-IGC for energetic heterogeneity determination helped in characterising the changes in surface properties of powders produced by different processing operations. Moreover, the experimental analysis combined with modelling approach and surface energy deconvolution were able to quantifiably determine different surface energy regions on the processed lactose samples.

508743 Nonlinear Radial Stress Response during Uniaxial Die Decompression

Abstract:

A nonlinear curve is generally observed at a plot of the radial stress versus axial stress during the entire loading/unloading procedure of a uniaxial die compression. At initial stages of unloading, the relationship is generally accepted to be linear. This observation serves as the basis of the argument that the compact exhibits linear elastic behavior during this period, and elastic constants such as Young's modulus and Poisson's ratio can be easily calculated. As such, the calculated elastic constants are further assumed to stand for the overall elastic constants of the powder compact. Here we propose, however, that the presumed linear relationship during the initial unloading is instead a transitional moment during which the die wall friction force changes direction. This is observed through analysis of compression in which axial force transmission and the coefficient of friction (CoF) are evaluated continuously during both the loading and unloading phases of compaction. Additionally, the effect of compact thickness on these parameters is studied, and observed to have a significant effect on force transmission but no impact on CoF values. This result supports the assertion that friction forces change direction when the compression cycle transitions from loading to unloading, which provides an alternative framework in which to understand the apparent profile linearity. The phenomenon is observed in materials exhibiting a broad array of mechanical properties. Using this analysis, we propose an alternative approach to determining fundamental elastic constants, in which an assumption of linear elastic strain was made to the overall (instead of a small portion) elastic recovery of a compact during unloading. With this new approach the elastic constants such as Young's modulus and Poisson's ratio can be calculated which is different from published data.

507759 Isolating Rate Processes in Spherical Agglomeration: An Investigation of the Breakage Mechanism Using a Contracting Nozzle

Abstract:

Spherical agglomeration is a size enlargement process in which crystals are agglomerated in-situ in a crystallizer through the addition of immiscible bridging liquid. This process has the potential to improve flowability, solubility and compactibility of crystalline drugs. It can be used for the production of tablets in pharmaceutical industry. However, research to date has been restricted to product specific investigation. We proposed a three-step process mechanism for spherical agglomeration: wetting and nucleation; growth and consolidation; attrition and breakage.

The aim of this study is to investigate the breakage mechanism in spherical agglomeration process by isolating the breakage rate process. Systems consisting of model particles and bridging liquids are used; plastic beads with kerosene as the bridging liquid and water as the liquid suspension. Bridging liquid drop size was controlled by using a microfluidic system. This enable a range of different drop/particle size ratio to be investigated.

Agglomerate breakage was studied by pumping a suspension of spherical agglomerate through a contracting nozzle placed in between two syringes. The particle sizes distribution was measured before and after the agglomerates were broken using Malvern Mastersizer 3000 laser diffraction. The effect of different velocity gradient on agglomerate breakage was studied by varying the diameter of nozzles and the applied flow rate.

Here we present the results from these novel experiments. Agglomerate size as a function of fluid velocity, constriction diameter, and number of passes is discussed.

Development Strategies of advanced bio-products based on particle design technology

Woo Sik Choi, Shigeki Toyama, and Won Hoon Park, KAPPIE, Busan, South Korea

Abstract

This paper deal with the establishment of "The Institute of Bio-High-Tech Prototype Development (IBHPD)" based on application of particle design technology (PDT) to promote the national activity. The advanced prototypes for various fields related to the Bio-High-Technology are included, that is, such final or mid-products and raw material powder as healthy functional foods including cosmetics and pharmaceuticals. The Institute accepts the proposal for the creative ideas from anyone due to the successful proposal for achievements of researchers from various fields. The validity and possibility of the proposal are evaluated by a determined feasibility test for the proposal is examined. The Institute decides to support the proposal if it is judged to be worthy not only Korean market but also the world market. The particle design technology (PDT) which is finally authenticated would be applied repeatedly. Simultaneously other fundamental technologies and regulations of Government are also examined thoroughly in advance.

1. Introduction

Every one equally wishes to be health. In other words, anyone has modest wishes to want to live a happy life in good spirits even growing old. According to the era of the trend, Bio-High-Tech researchers in University have raised a lot of achievement by their sincere devotion to basic and application researches. However, such achievements have not always created effective biological products, but they spend time indispensable.

As one of the solution to this, we propose here an effective vision to install and operate the Institute of Bio-High-Tech Prototype Development (IBHPD) under KAPPIE. The IBHPD recruits various ingenious and the creative idea with their study results (data or discovery). Then the committee of IBHPD executes the validity or feasibility for commercialization of the proposal.

If the results of its review on a proposal are recognized to be a valuable, the bio-products can be prevailed in Korea or in the world. The committee of IBHPD selects the proposal as a prototype of the bio-prototype development project and then supports the development of the prototype. After confirming the efficacy and safety for the subsided prototype, the bio-prototype will be recognized as a succeeded bio-prototype.

This Institute plays a role to exert fully elemental technologies for each step, to minimize the side effects, and to maximize the efficacy of bio-products, namely healthy functional food, cosmetics, and medicine etc. in order to newly create novel bio-products. The prescription will be derived by experts of Western medicine and Oriental medicine as well as experts of particle design technology. As a result of the normal activity of the Institute, Korea holds 5% of the world market of bio-industry (the world market in 2010 of $1560 billion, the domestic market of $9.0 billion). Through the world-class market of bio-products by creating the 10 WP-KBM (World Premium Korean Bio-Medicals) and by developing the related products, it will be believed that Korea can be greatly extended to occupy more than 16% of the world market within 10 years. [1, 2] Herein, The 10 WP-KBM has been named according to the WPM, World Premium Materials, 2010 Materials Development Program of the Ministry of Knowledge Economy, Korea at that time.

The establishment of the Institute has been appealed to the government in every opportunity [3, 4] and to academic societies by presentation [5, 6]. In addition, I would like to let the members of KAPPIE have the pride that they can contribute to the national development through bio-products, industrial materials and joins technology development to the members who actively participate in the installation of the practical operation of test laboratory for the bio-prototype.

II. Present state of the Korean medicine biotechnology industry

The hybrid of eastern medicine and western medicine will take the advantage of the ancient traditional oriental medicine as an important axis for the development of advanced bio-prototype. The Ministry of Health and Welfare of Korea has invested a total of 1 trillion Won by 2015, and announced to foster 10 trillion Won Korean medicine industrial markets. Although Government has supported the herbal medicine sector but it was insufficient so far in 2011. Then, five-year investment of about 1.0099 trillion won was planned in 2015. This will be well matched with the partial objectives of the Institute. Some topics are discussed at the presentation site

From 'Worldwide Prescription Drug & OTC Sales by Technology (2008-2022)' published by EvalutePharma in 2017, the uptake of biologics is expected to continue to keep at same level from novel biologic blockbusters entering the pharmaceutical market. The penetration of biotech products is set to increase from a 25% market share in 2016 to 30% in 2022. According to the latest reliable forecast in 2022, 52% of the top 100 products will come from not chemical products drop off the patent cliff but breakthrough biologics get approved. [7]

III. Overview of the Institute of Bio-High-Tech Prototype Development (IBHPD)

3.1 Major development contents

(1) Integration manual of PDT and element technology for the bio-prototype development: Particles design method for development of the novel functional materials created through the sophisticated combination of various materials (inorganic, organic, metallic) and the control of particle size distribution and morphology of materials (large particles, medium particles, and small particles, the particle size of ultrafine particles), and the optimized operation units such as milling, granulation, and compaction are utilized by better combination with the manual set in integrated technologies.

Professor Koishi contributed an article entitled "Polyhedron-like particle Design Technology for the Advanced Functional Fabricating of Fine Particles" in the special issue about the composite and function of the particles and described as follows: For example, the technology road maps of APPIE's Technical Division activities, especially granulation technical division, fine nanotechnology technical division, battery manufacturing technology technical division, static electricity utilization technology technical division, wet process technical division, mixing and molding technical division, and grinding technical division are greatly utilized for this Institute. Future technologies are described in the technology road map of each technical division. It is considered to be essential to perfect understanding and integrated usage for particle design technology [10]. He has published a practical guide book for PDT that "the compass-like hint and thinking idea" makes the best powder material for making the final form, where the creation of the morphology to ultimate powder material is composed of four categories.

After completing successfully the first step, the institute advances to the next step, construction and practical training of joint technology support system of Industry-Academy-Government, to the specialized step, specialization of element technology and bio-technology process, and finally the last step to create the novel bio-prototype.

The contents are to make the particle shape of the nano scale level, to create an integrated powder form, to create non-familiar form of learning from nature, and to challenge to the new industrial change [11]. The key point is the territory where is capable to expand this idea to particle size.

3.2 Establishment strategy

The basic concept and management solution of the Institute of Bio-High-Tech Prototype Development are briefly summarized in Scheme 3.1.

Scheme 3.1. The basic concept and management solution of IBHPD

Step	Main activities	Important activities	Authority
1st	Receipt of bio-prototypes	- Seeds providers (Univ., Institutions, and general public),	Proposer
	Proposal Review	- Needs consumers (Industry, Public authorities),	KAPPIE
		- Consultation of proposals: efficiency, reliability, Timeliness, Economics, Technology,	BPDC [1]
2nd	Prototype production	- Bio-prototype test lab. (in member's company)	Member
	Allocating manuf. TL	- Particle Processing Technology 2	BPDC
	Complement need	- Complement bio University	Ind.-Univ. coop.
	Meet challenges	- PD of main ingredient and raw materials	Test lab.
		- Manufacturing bio-prototype by unit operations	Test lab.
3rd	Evaluation of prototype	- Primary efficacy and safety assessment	ISC [3]
	Conf. of its characteristics	- Repeat steps until reaching final goal	
4th	Completed prototype	- Cooperation with related institutions of final deliberations	KAPPIE
	Final tech. transfer	- Establishment of PDT confirmed for Bi-products	
		- All contracts signed for know-how to customers	

1 Refer to the configuration of the prototype development committees.

2 The advisory and technical road maps of APPIE technical divisions, Japan.

3 Install and operation of SMEs' Integrated Support Center (University)

3.3 Organization

The Institute is organized, established and managed as one of the think tanks in KAPPIE. In accordance with the spirit of the Institute, each member installs and operates a special prototype test laboratory with his Know-How inside his company. KAPPIE establishes and manages a bio-prototype development ex-pert committee for consultation between the relevant authorities.

The Bio-prototype test laboratory (Test Lab.), which is installed and operated by each member company is determined simultaneously when the establishment from Government. The decision for establishment of the test labs. is made by the public offering or recommendation of the expert committee with deep understanding of the spirit of the Institute. Tentative items are shown in Scheme 2.2. They are the basic powder technology type of the test Lab., the purpose-oriented technology type of the test lab., and the next-generation technology of the test lab. The companies specially to participate in establishment of a specific test laboratory should receive special incentives.

The bio-prototype development special committee will be composed of 7 to 9 experts of the relevant institutions including a chairman. KAPPIE convenes expert meeting of the bio-prototype development special Committee according to a request of proponents. The committee deliberates important issues and makes decisions with the cooperation to achieve the best ways to facilitate the completion of the bio-prototype.

Scheme 2.2 shows the major components of the Institute and tits outline of the operating method. The key point is to carry out the validity of the bio-prototype based on the particle design technology. According to validation data, and the test labs. of members are to be assigned in charge of the test laboratory bio-prototype for each manufacturing process. The bio-prototype integrated support centers for small-medium companies installed in the University will serve to evaluate the efficacy and safety of the bio-prototype. Through a series of processes mentioned above, the more outstanding bio-prototype will be created.

Scheme 3.2. Major components of the Institute and its outline of the operating method.

1. Bio-tech prototype professional development committee of 7 to 9 persons.
2. Special test laboratories installed in member's companies themselves

Basic powder technology	**Purpose-oriented technology**	**Next generation-oriented**
Unit operations such as	Medicine, Food, Cosmetics	Energy, Traditional tech.
Grinding, Granulation, etc.	High value-added products, etc.	3D Printing tech., AI, etc.

3. Integrated support center of bio-prototype development for small-medium business in University1)

1. Role of ISC: 1) Measurement support for basic and applied physico-chemical properties of bio-prototype, 2) Efficacy validation support of bio-prototype, and 3) Safety assessment support of bio-prototype.

2.4 Establishment site of the Institute and its networks of the Institute

The Institute is installed and operated as an activity of KAPPIE as a think tank. In order to promote smooth future cooperation with other research institutions such as University, the branch of the Institute is recommended to configure in the related Universities.

The specialized committee of bio-prototype development in Association and the network of similar institutions are connected on each website and the useful information is also disclosed in principle.

IV. Sources of Seeds and consideration on its utilization

Table 4-1 summarizes the sources of health (functional) foodstuffs prototype including raw materials of medicines and cosmetics prototypes.

Table 4-1. The sources of Seeds for the development of healthy functional foods bio-prototypes, raw materials for medicine, and cosmetics bio-prototypes.

1. Basic and applied research papers and practical or dormant patents from various institutes
2. Cooking notes on the recipes of traditional local products and their tasting methods
3. Prescriptions of many traditional oriental herbal medicines, for example, Sibjeondae-tang
4. Comment of citizen on bio-products in traditional markets and large supermarkets.

Table 4-2 shows a summary on Sibjeondaebo-tang composed of 10 kinds of Chinese herbal medicines and its related data such as Prescription related with Chinese name, Latin name, Main ingredient, Ratio, Part of plant, and Pharmacological action.

As a typical example for particle design, the reason why Sibjeondaebo-tang is selected is composed of two Chinese herbal prescriptions composed of 4 kinds of Chinese herbal medicines including two Chinese herbals. Sibjeondae-tang is willingly used when lacking in vitality or energy and coldness of body are felt. The main efficacy is widely used as a prescription used to protect the body when the condition is poor due to such decrease as physical strength after illness, fatigue, and anorexia, etc. [13-22]

Table 4-2 also suggests many subjects capable to apply a particle designer for selection of material and processing technique as follows: 1) the main chemical components and major actions in each herbal medicine, 2) through examination with the pharmacological chemical composition of the active drugs used in present medical formulation, 3) preclinical efficacy examination when some herbals are added in or removed from prescription of Sibjeondaebo-tang, and 4) basic research for the existing and unidentified active ingredients in medicinal plants as well as the useful treatment processing technology.

It is very important for the perspective bio-product creation as well as establishment of particle design technology to examine the formulation of present used medicines, its ingredients, its contents, and its dose with the important pharmacological action of each Chinese herbal medicine including classical and experimental data.

It is also suggested that the use of recently developing AI robot will be a great help in the near future for the breakthrough of this field. The close collaboration between various interdisciplinary researchers is sincerely requested. The fields include chemical engineers and others such as basic researches on standardization of traditional Chinese medicines, the compositional structure of medicinal herbal plants, and their pharmacological actions [21,22].

Table 4-2. Summary on Sibjeondaebo-tang composed of 10 kinds of Chinese herbal medicines and its related data

Prescription	Chinese & Latin name	Main ingredient	Ratio(g)*	Part	Pharmacological action
Sagunja-tang	人蔘 PANAX GINSENG C.A. Meyer	Ginsenoides Rg1, Rb1	3.75	Root	Warm body & metabo.
	白朮 ATRACTYLODIS RHIZOMA ALBA	Atractylon, Acetoxyatractylon	3.75	Roots	Water's uneven distrib.
	茯苓 PORIA SCLEROTIUM	Pachymic acid, Triterpenoid comp.	3.75	Sclero.	Diuretic inhibiting
	甘草 GLYCYRRHIZAE RADIX	Liquiritin, Glycyrrhizin	3.75	Root	Action relieving tension
Samul-tang	地黃 REHMANNIAE RADIX	5-hydroxymethyl-2-furaldehyde	3..75	Root	Control of the blood
	芍藥 PAEONIAE RADIX	Albiflorin, Paeonniflorin	3.75	Root	Relieving muscles&vessels
	川芎 CNIDII RHIZOMA	Ferulic acid	3.75	Roots	Stimulating blood circul.
	當歸 ANGELICAE GIGANTIS RADIX	Nodakenin	3.75	Roots	Gynecologicals Improve.
Other	黃耆 ASTRAGALI RADIX	Isoflavone, l-canavanine	3.75	Root	Water disorder of body
Other	桂皮 CINNAMOMI CORTEX	Cinnamic aldehyde oinnamic acid	3.75	Bark	Moving of stagnating

*1 Mixing ratio is referred to prescription of 'Doguibogam' as the reference from the book: Korea Institute of Oriental Medicine, (2013), "Standard Chinese herbal Prescription Drug information 2013", pp. 223-257. [13]

2 Sibjeondaebo-tang has important pharmacological functions or effects: Bacteria proliferation suppression [11], Tumor formation and Transition suppression [12-50], Reduced immune response, hematopoietic function recovery, and earl to re-action inhibition [51-84], Neuroglia of oxidative damage defense effects and Alzheimer's dementia prevention and protection [85-87], Blood clotting inhibition, creation of blood clots and prevent [88], No liver damage, Reduction of ammonia in blood, and liver metabolic deficiency recovery [89-91], Healing and regenerative of skin wounds [92-94], Radiation protective effect [96], Increase of iron bioavailability, No change of heavy metal of its concentration in blood and antioxidant properties [97-81]. Reference Number in [Bracket] of each function are as reference of above number (1).

3 Sagunja-tang: Prescription is composed of 4 kinds of Chinese herbal medicines such as Panax ginseng C.A. Meyer,

Atractylodis rhzoma alba, Poria sclerotium, and Glycyrrhize radix. The prescription is used in improvement of gastrointestinal function and congestion of water, and also for easily exhausted man.

4 Samul-tang: Prescription is composed of 4 kinds of Chinese herbal medicines such as Rehmanniae radix, Paeoniae radix, Cmidii rhizome, and Angelicae gigantis radix. The prescription makes the circulation of blood flow better and the body warm. It is good prescription for woman because of smooth skin and hormone balance effect.

5 Palmul-tang: Prescription is composed of the summation of 4 kinds of Chinese herbal medicines of Sagunja-tang and 4 kinds of Chinese herbal medicines of Samul-tang. The prescription enhances the metabolism such as fatigue, anemia loss of appetite, dizziness, and discharging of old shit.

V. Expected results and future development plans

The particles design technology confirmed by this Institute with the developed process of bio-prototype is expected to be contribute to the high technology products such as the development of other biotechnology products and industrial materials. Furthermore, small and medium-sized companies having no regular lab-oratory are capable to have a pride to contribute to country's development, and to new industrial materials development and high quality of existing products. Furthermore, the technology development for high added value products is widely expected with synergistic effects on the creation of brand technology products to meet the global standard.

Moreover, by establishing a processing technology for powder material applied to 3D printing technology that is met great attention as a technique for leading the next generation, the business entry in this field is greatly expected. In addition, it is expected that more young founders will be cut the tape of their businesses.

The industrialization of basic research by matching the Seeds and the Needs is expected to make the biotechnology expand into other industrial material industry. [22].

VI. Conclusion

This Institute should promote the research and development with intensive investment for the creation of advanced bio-prototype by applying the preparation technique of the powder material, the composite particles, and the functional particles,

In order to achieve this Institute smoothly, the joint research projects between this Institute and the research support institutions under the Ministry of Education, the Ministry of Health and Welfare, and the Ministry of Industry Trade Natural Resources should be tightly connected for the efficient use of national budget. The 10 items of bio-prototype development plan including some idea for the establishment and operation of the Institute during consideration are shown in Appendix 1 and useful references.

When providing the Seeds of prototype, this appendix will be helpful. Since our nation of underlying strength is excellent, a number of Seeds proposals are sustained.

In addition, the help of craftsmen that can be willing to process successfully the proposed Seeds would also be abundant.

References

1. http://www.motie.go.kr/motie/ne/rt/press/bbs/bbsView.do?bbs_seq_n=64157&bbs_cd_n=16, (2013), WPM Project.
2. WS Choi (2015), "10 WP-KBTM (World Premium Korean Bio-Particle Materials) on presentation".

3. WS Choi (2003, 2009), "Technical Report on Powder Technology".
4. WS Choi (2015), "Proposal to MOTIR for Institute of Bio-Prototypes of Industrial Field".
5. WS Choi (2007), "Activity and Prospect of the Korea Association of Powder Process Industry and Engineering (KAPPIE)', Proceedings of the 1st Asian Symposium of Pharmaceutical Sciences and Technology at Shenyang, China.
6. WS Choi et al., (2009), "Activity and Prospect of the Korea Association of Powder Process Industry and Engineering (KAPPIE) and some members: a basic idea for development of novel advanced raw materials and parts", Proceedings of at the 4th Asian Symposium of Powder Sciences and Technology, New Delhi, India.
7. EvalutePharma, (2017), "World Preview 2017, Outlook to 2022", p. 12.
8. BH Han (2008), "Support Methods of Bio-Tech-Products", Promotion Institute of Korea Health.
9. http://www.hw.go.kr/ (2011), "The second Development Plan for Korean Traditional Medicine and Education from 2011 to 2015".
10. http://www.appie.or.jp/ (2014), "Technical Division of APPIE".
11. Masumi Koishi (2013), "Polyhedron-like particle Design Technology for the Advanced Functional Fabricating of Fine Particles", Funtai Gijutsu, 5(5), 411-421.
12. C. Buglistello (1968), "Health Care Needs and Prototype Plans Technology-Health System, Report of a study sponsored by the Committee on Interplay of Engineering with Biology and Medicine of the National Academy of Engineering", Carnegie-Mellon University, October 1.
13. Korea Institute of Oriental Medicine, (2013), "Standard Chinese herbal Prescription Drug information 2013", pp. 223-257.
14. H.Y. Lee, H.K. Ha, D.Y. Jung, N.H. Lee, and H.K. Shin, (2010), "Immuno stimulatory activities of Samul-tang, Sagunja-tang, Pamul-tang and Sipjeondaebo-tang in vitro", J. of Oriental Neuropsychiatry, 21(4), 41-51.
15. K.G. Lee,·D.S. Huang, Y.B. Yu, J. Y. Ma, H.K. Ha, and H.K. Shin, 92010), "A Study On Compositions, Dosages, and Usages of Sagunjatang, Samultang, Palmultang, and Sipjeondaebotang in Literature", J. of Oriental Neuropsychiatry, 21(4), 41-51.
16. B.H. Jeon, W.H. Woo, H.B. Moon, (1999), "Effectof Sagunjatang Extract on Cytokines Secretion in the Cultured C", Korean J. Oriental Medical Pathology, 13(2), 84~90.
17. Y.S. Kim, Y.H. Lim, S.G. Wang, S.J. Yun, and C.H. Park, (1999), "The Physico-chemical Properties and Anti-oxidation Effect of Samul Chol-Pyon", J. Korean Soc. Food Sci. Nutr., 28(5), 990-996.
18. H.J. Park and W.S. Ko, (1998), "The Experimental Effects of Palmul-tang, on Anti-Cancer and Immunologic Function", K.O.M.S., 19(1), 327-338.
19. http://www.jah.ne.jp/~kakp, (1996), "Home page of No. 110 of medicine: Well-understanding of Drug from Hospital".
20. http://www.tsumura.co.jp/english/kampo/ About Oriental medicine of Tsumura Co., Ltd.
21. Hong-zhou Wu, Zhao-qin Fang, and Pan-ji Cheng, (2013), "Introduction to Diagnosis in Traditional Chinese Medicine", pp. 3-65, World Century.
22. Kim Sung-Hoon, Kim In-Rak, (1997), "Brief Introduction of Traditional Chinese Medicine (TCM) and Possible Measures to Protect the Endangered Species of TCM", The Journal of Korean Medicine, 18(2), 366-373.

Appendix 1. 10 WP-KBTM Prototypes from the view point of World Premium Korean Bio-Particle Materials.

I. Long-term goal of ultimate development bio-products

1. Cold medicine
2. Renewal of the existing Chinese medicine prescription bio-products
3. New lighting of the lifestyle-related disease
4. An increase of immunity and prescription of the intractable diseases
5. Novel formulation in conformity with oriental references including 'Dongeuibogam'

II. Short-term development bio-products for healthy function foods and cosmetics

6. Prescription of nutritional supplements for the high aging society
7. Prescription of the healthy functional food based on with oriental references including 'Dongeuibogam', Handbook of Oriental Medicine and Therapy.
8. Globalization of the Korean traditional bio-products, for examples, Red Ginseng, garlic, and fermented foods*.
9. Prescription of Chinese herbal medicine cosmetics for high aging person
10. Prescription of Chinese herbal medicine cosmetics for the woman

*1 Particularly, there are the local unique healthy foods in country or all the fields. The Institute selects it as Seeds of bio-prototype and tries to develop it a bio-prototype for healthy functional foods by using Korean traditional techniques and recent particle design technology established in this Institute. And then the Government supports the industrialization of the bio-prototype for Seeds provider. When the provider of Seeds is a foreigner, the Government fund supports the industrialization of the bio-prototype for the health and welfare of the country as an international cooperation project between Korea and the country.

Healing formulation of South Korean loess powder as a bio material and its application to environmental friendly bio products

Woo Sik Choi, Shigeki Toyama, KAPPIE, Busan, South Korea
Yeon Jin Choi, Hanul Bio Co., Yangsan, South Korea

Abstract

Loess mined in Korea, especially yellow loess has been inherited as a mysterious soli to greatly contribute not only health promotion but also environmental clean-up from long years ago. When this powder was applied to various skin diseases, many prominent healing effects are observed such as atopy, face pack, half bath and full bathing, frostbite, and athlete's foot. And also many healing cases were reported from many users, for example, rest sleep, restoration of fatigue, and relieving pain and stress were reported from many users. The paper aims to discuss the healing properties required to the eco-friendly bio products.

1. Introduction

Loess powder produced from South Korea has been known to have contributed to the basic healing for health maintenance. The loess powder contains various mineral components, and has a large specific surface area. It has long been said that loess powder indicates an inherent excellent far-infrared radiation function such as uniform heating and heat storage effect, deodorant, excellent antimicrobial and mold prevention effect, temperature and humidity control function, and natural cosmetic effect. [1,2] Yellow loess powder has long been used in not only making loess bowl but also building a house. Loess's house is cool in day and warm in the night. Its warmth brings about recovery from fatigue.

The United States and Russia have studied human survival conditions under the harsh environment of the universe through the space development project that began from the 1950s. In particular, the research on solar radiation of the National Aeronautics and Space Administration (NASA) confirmed that far infrared rays with wavelength ranging from 4.0 to 14.0 μm are indispensable for the survival of organisms. [3]

German scientists who have studied bio-resonance at the simultaneous age attempted to associate it with the mechanism of human health recovery as a cure healing method so-called natural therapy [4]. In Japan, far-infrared rays have been attracted attention from health and energy saving fields, and various ceramic materials were developed as the far-infrared oscillator. [5-7]

Hanul Korea Co. has developed a process to manufacture loess ball indicating the radiation characteristics of far-infrared rays' emissivity of 92.7%. [8-12] A series of studies have been conducted to confirm the effectiveness and safety of loess bio-ball bio-products from the next three items: (1) Measurement of fundamental physico-chemical properties of loess, observation of red blood cell movement during viable blood by Somatoscope microscope, observation of DITI photograph (digital infrared thermal image data) used for medical diagnosis, (2) Discussion referring to oriental medicine literatures including 'Donguibogam', an old oriental medicine handbook and recent research papers, and (3) Statistical analysis of case reports experienced over the years by many users who have been used loess ball products in connection with such environmental factors as detoxification, deodorization, air cleaning, relief of red tide, antifungal effects, and recovery of fatigue and cure of stress relief by temperature control, etc. [13]

This study aims to find out the action mechanism including the healing characteristics to be provided as a bio raw material powder in the future by clarifying the various healing properties for recently developed loess ball bio-products, especially S4 bed.

2. Experimental method

2-1. Measurement of physico-chemical properties of loess powder

Qualitative analysis of XRD by Rigaku miniflex 600, measurement of particle size distribution by MT 3000 II, MICROTRACK, evaluation of particle dispersibility by Lumisizer 651, observation of surface morphology by SEM and absorption to desorption behavior by equipment were carried out. The obtained physical and chemical characteristics of Korean loess powder are summarized in Table form. [14, 15]

2-2. FT-IR spectrum measurement

Far-Infrared rays absorption spectrum and FT-IR were measured to evaluate the far-infrared rays' radiation properties. [15, 16]

2-3. Living blood test and Digital Infrared Thermal Image (DITI) inspection test

Two experimental methods were adapted to evaluate the preclinical efficacy of loess bio-ball. One is the living blood test by somatoscope and the other is DITI inspection using IRIS-XP by thermal diagnostic device. [17]

2-4. Summary of reports for healing cases experienced by users

Although it is required to confirm the therapeutic effect by clinical tests of the loess or the loess bio-ball products, the enforcement has been delayed due to various circumstances. A possible method at a moment is to have a frank conversation with the user who has experienced the healing effects of loess bio-ball products. It is still worthy to utilize the information reports for the degree of change of the body symptoms from users before and after using the bio-products. So, a number of reports of healing cases experienced by the users of loess bio-ball products are summarized. [13]

2-5 Manufacturing process of loess bio-ball based on PDT

The loess bio-balls are manufactured by a series of unit operations as a low temperature drying process by using selected loess ball. [8, 9]

3. Results and Discussion

3-1. Measurement result of physical and chemical properties of loess powder

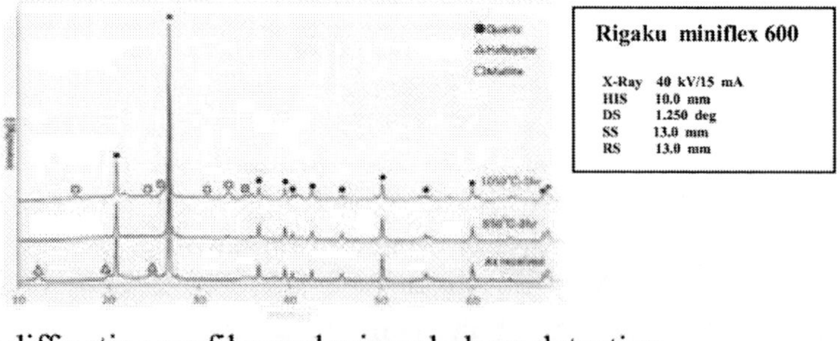

Fig. 3-1. X-rays diffraction profiles and mineral phase detection in various heat treating processes of Korean loess by Rigaku miniflex 600.

Figs. 3-1 shows the qualitative analysis results of XRD for sample of two treating process. The measurement conditions are as shown in right of the figure. The XRD pattern of the powder samples prepared by various heat treating processes was shown together with that of the raw loess powder. These patterns indicate that the supernatant powder subjects to water elutriation process was halloysite ($Al_2SiO_2(OH)_4$) and the precipitated powder was quartz. [15]

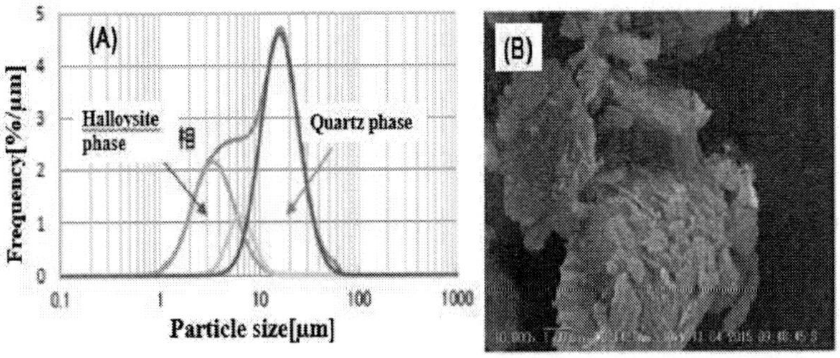

Fig. 3-2A and B. Particle size distribution (A) and particle shape (B) of loess powder.

Fig. 3-2A and B show the measured particle size distribution of loess powder and the photograph of particle shape. Three frequency distributions obtained by dividing the original distribution by peak separation technique based on the assumption of a log-normal distribution function are shown together under the curve of the particle size distribution (A). Loess is composed of quartz phase particles and halloysite phase particles, and the halloysite phase may be in the secondary form of agglomerated particles. Fig. 3-2B shows the shape of loess particle measured by microscope. Samples were dispersed in ion exchanging water with 0.5 wt% sodium hexameta-phosphate under ultrasonic by MT3000 II, MICROTRACK. [15]

Fig. 3-3 is a plot of the heating curve such as a series of adsorption and adhesion process. The BET's specific surface area is 61.65 m2/g for powder and 25.466 m2/g for granule, indicating type II adsorption isotherm with H3 type hysteresis. The absorption amount of water vapor is 69.8 cm^3 (STP)/g for powder and 48.6 cm^3 (STP)/g for granule. Although the porous properties of the loess particles are not so high, some slit-like voids between the plate-like particles are observed. The characteristics of loess powder and granule are summarized in Table 3.2. [15]

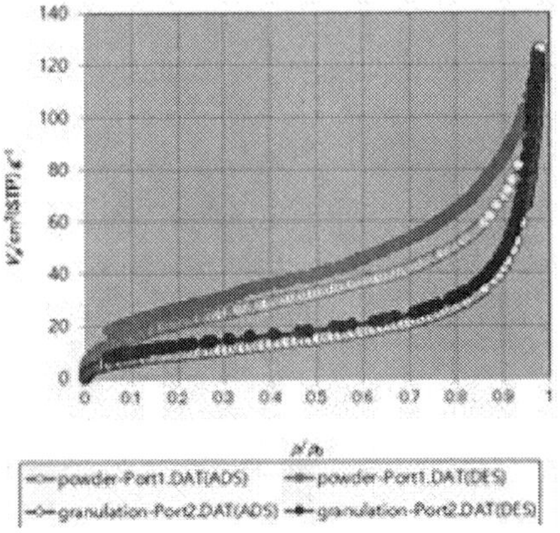

Fig. 3-3. Absorption & Adsorption isothermal curve.

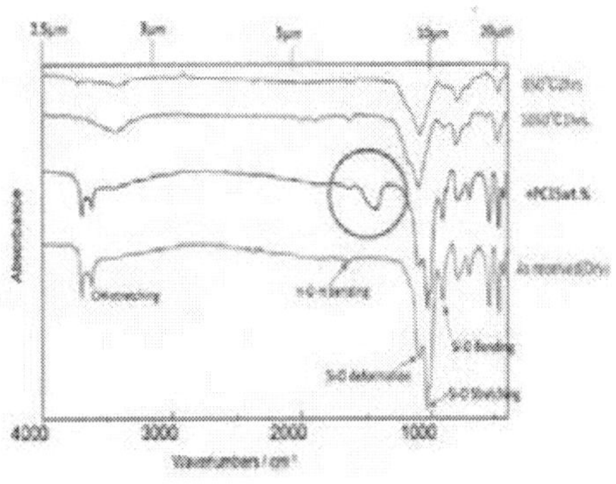

Fig. 3-4. IR absorp. Spectrum vs. wave number.

3-2 Far-infrared radiation rays characteristics of loess particles

The far infrared emissivity of loess powder was measured at 40°C by the FT-IR spectrometer at the Korea Building Material Test Institute. The results indicated that the emissivity was 0.927 and the radiation power was 374 W/(m2·μm) at 40°C. [15]

Fig. 3-4 shows the IR absorption spectrum of Korean loess samples treated from various treatment methods as shown in figure, where two meaningful peaks are observed. They are related to be nitric oxide action as a signaling molecule in the cardiovascular system in NIR wavelength at the Nobel Prize in Physiology or Medicine 1998. [3,4], and to thermal heating and detoxification by resonance vibration of water molecules in 9.5 μm. [1,2]

Table 3-2. Summary of characteristics of absorption and adsorption isothermal curve for Korean loess powder and granule of bio-balls.

Parameter	Unit	Powder	Granule
V_m	[cm^3(STP) g^{-1}]	18.358	7.5827
BET	[m^2 g^{-1}]	61.654	25.466
Total pore volume (p/p_0=0.980)	[cm^3 g^{-1}]	0.094572	0.1016
Average pore diameter	[nm]	6.1357	15.962

3.3 Preclinical examination of living blood test and DITI inspection test

Man of 50's generation Woman of 50's generation

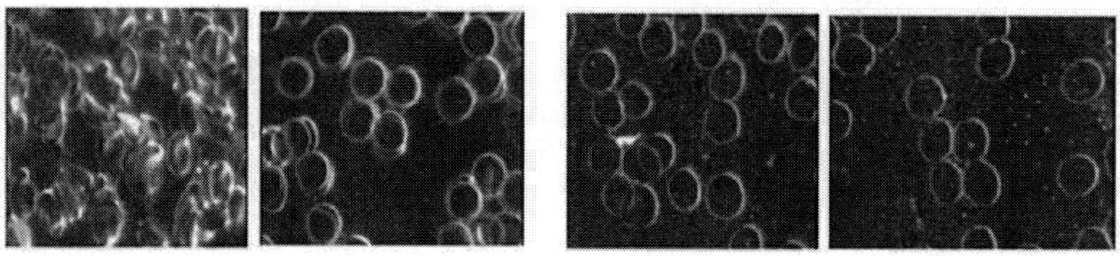

Left: before radiation, Right: after 20 minutes of radiation on bio-ball mat.

Fig. 3-5. Somatoscope photographs for living blood test of two persons.
Experimental equipment: Somatoscope: Olympus BX50, CCD camera: Toshiba, Tokyo, Japan.

Fig. 3-5 shows the Somatoscope photographs for living blood test of two persons under experimental conditions as shown in figure. From the Somatoscope photographs, we can find out that the red cells in living blood after 20 minutes of radiation on the mattresses are separated independently one by one and move more actively than those of before radiation. The blood circulation after 20 minutes exposed to radiation on the mattresses is observed to be improved not only in a large blood vessel but also in a peripheral small blood vessel. This movement of red cells in the case of the mattresses at room temperature is also observed to be the same movement of red blood cells from those photographs. [9,12]

DITI is a diagnostic technique that is non-invasive and no exposure from radiation rays. This technique is employed in common clinical uses such as early detection of breast cancer, monitoring changes in overall health, monitoring healing processes, disease and virus monitoring, and fever screening (i.e. H1N1, SARS) based on color changes before and after treatment, which indicate the influence of increases or decreases in infrared radiation emitted from the body surface. [17]

(A) Before expose (B) Expose from loess bio-ball products (C) Expose from common loess soil products

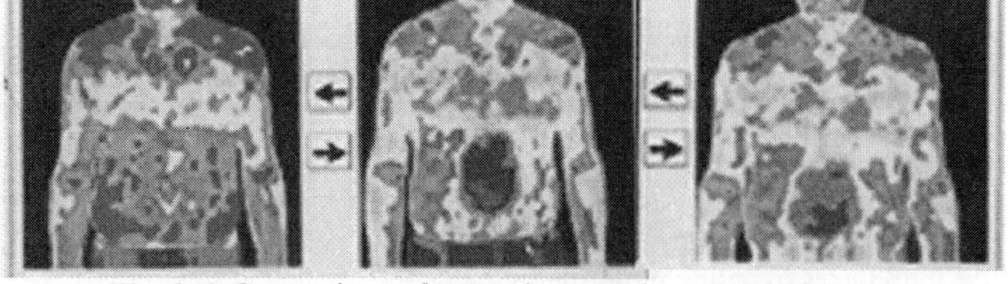

Fig. 3-6. Comparison of DITI photographs at (A) before expose, (B) exposed from bio-ball products, and (C) exposed from common loess soil products.

3.4 Application of loess powder to health

Raw loess was aged for 6 months or longer, and then it was stabilized by removing moisture and foreign matter such as small stone and roots of plant. After that, it was disintegrated or ground. By a classification step, a fine powder with a good dispersion state was formulated according to the application usage. Except for the molding and drying processes mentioned in the previous reports, it is almost the same processes. [8,9] However, the obtained loess of fine powder was divided into small pieces, and its effectiveness and safety were investigated with the rule of quality control.

Since loess powder has long been said from ancient times to be excellent in detoxifying power, adsorptive power, self-purification ability as described in above introduction, it shows also a peculiar effect on the human skin. Thereby removing waste matter in the skin, it has been said that it works well for the prevention of skin ageing. [12] Here, we reported surveyed some details examples of experienced cases that could confirm the curative effect in references. [13].

3.5 Usage of loess powder and healing effect

It is summarized how to use loess bio-ball products from users and many healing sites.

1) Usage as a half bath and bath agent: Add hot water of 30 to 40°C to the bath and disperse 100 g of raw loess powder. After soaking for 10 to 20 minutes and sweating, body was washed with warm water. Then, such improvements were observed as skin rash, smooth skin, alleviation of acne, and pimples of arm.

2) How to use it for atopic dermatitis: Disperse 100 g of powder in water, knead with cream and apply with brush or hand. After 10 to 20 minutes, wash it thoroughly with warm water. And then the relaxation and improvement effect of atopic dermatitis are felt.

3) How to use face pack: Add 10 ml of water to about 10 g of powder and knead it, spread it on your face with a brush or hand and paint it. When the powder dries and feels tight after 15 to 20 minutes, wash with warm water and finish with cold water. Instead of water when packing, milk or yogurt etc. mixed according to user's preference can be used.

4. Consideration

4-1. Investigation research for Korean general loess-room

Prof. Y.K. Kim in College of Nursing, Catholic University of Pusan studied on a subject about 'perceived healing effects after using Korean loess-room'. [23] The loess used in this loess-room is a common loess powder as shown in Fig. 3-6C and is different from the loess powder in this study. The research results are summarized with 5 theme and its some categories. Namely, the promotion of circulation of blood is 1st theme. The relaxation of strained mind is 2nd theme. Being refreshed is 3rd theme, Skin being kept moistly is 4th theme, and body waste being eliminated well is 5th theme. 41.35% of subjects benefited from the disease or disorder, 58.65% of subjects did not yet judge, and no one of subjects felt the side effects. [18]

4.2 Heat balance equation of human body

The heat balance equation of the human body describes the heat balance between the human body and its surrounding thermal environment. The specific data related with the body and the environment is required to calculate this body heat balance. It is tried to describe body's heat balance equation with environment in loess-room.

A general heat balance equation between the body and the environment is described as following Eq.:

$$\mathbf{S = M \pm W \pm R \pm C \pm K - E - RES} \qquad (4\text{-}1)$$

where S is rate of heat storage, M rate of metabolic heat production, W rate of mechanical work accomplished, R rate of heat exchange by radiation, C rate of heat exchange by convection, K rate of heat exchange by conduction, E rate of heat exchange by evaporation, and RES rate of heat exchange by respiration in unit of [W/m2].

Six factors for above Eq. are related with air temperature, radiant temperature, and air velocity from environment side and humidity, clothing, and activity in human side. From Fanger's thermal comfort model and ISO No. 7773 as a trial health index of the loess-room, the values of PMV calculated at 4 heating conditions such radiant temperature of lower section of S4 bed of 20°C, 28°C, 33°C, and 38°C are shown in Table 4-1. Some users who want to sweat have willingly tried the high temperature of even above 38°C of lower section in S4 bed. [19, 20]

Table 4-1. Trial calculation of PMV for S4 bed calculated by Fanger's thermal comfort model

Run	1	2	3	4
M (W/m^2)	115	115	115	115
W (W/m^2)	0	0	0	0
Ta (°C)	20	20	20	20
Tr (°C)	20	28	33	38
V (m/s)	0.1	0.1	0.1	0.1
RH (%)	50	50	50	50
Icl (clo)	1.0	1.0	1.0	1.0
PMV	0.68	0.54	1.10	1.68
PPD (%)	6.9	11.1	30.5	60.9

4.3 Healing effect of loess bio-ball and its products

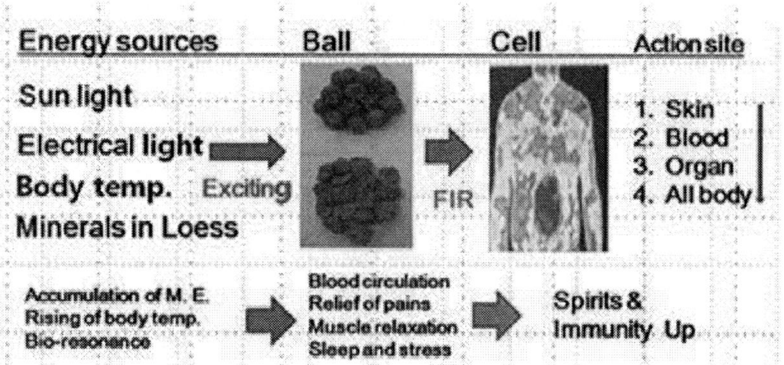

Fig. 4-1. Estimation mechanism on healing action of loess bio-ball based on experienced customers' reports of users, research of FIR, and related literatures.
*The thermal radiation of infra-red rays discovered by a scientist, William Herschel in 1800 has been used effectively for millennia to treat / ease certain maladies and discomforts.

Based on the results of the previous studies, the following several points are notified. (1) Measurement of fundamental physico-chemical properties of loess, observation of active red blood cell movement in living blood by Somatoscope microscope and of rising skin temperature in DITI photos, (2) Survey of Chinese medicine literatures including 'Donguibogam' and recent research papers, (3) Consideration of detoxification, deodorization, air cleaning, anti-fungal effect, and (4) Preclinical examination of healing effects which many users felt fatigue recovery and stress relief cure over a short period of time through summary of experience cases of users' reports by using loess bio-ball products. The mechanism of healing action us estimated as summarized in Fig. 4-1 [21-23].

4-4. Applicable products of environmentally friendly bio-ball

Although there is no definite medical evidence yet for bio-ball products, many user wishes to recover the health as shown in experience reports of many users on pain, stress, and headaches, etc. as a cure for bio products such as pillows, bed mattresses, cushions, and so on.

Summary of user's experienced reports based on major symptoms are summarized before use, and well-healed physical condition after use for S4 beds from 2014 to 2015 are listed up as follows: 1) skin's moist feeling and itching of the skin, 2) relief signs of head, shoulder, waist, and specially caliber pain symptom, 3) refreshing life through healthy recovery, 4) sleeping easily, and relieving insomnia, and 5) relieving stress by relaxing all mind and body are summarized.

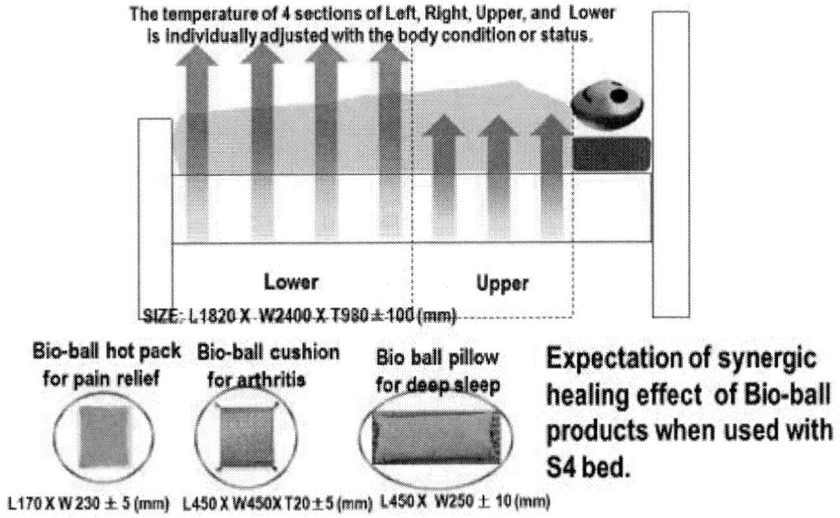

Scheme 4-1. Expected healing effects of loess bio-ball estimated from previous research results for the S4 bed

The S4 bed is designed by applying Korean traditional 'ondol' heating system and in S4 bed the temperature of the part divided by the left side, the right side, the upper side, and the lower side of the bed surface divided into 4 sections, can be adjusted independently according to user's physical condition and body status. If bio-products such as bio pillows, bio hot pack, and bio cushions are simultaneously used with S4 bed, some high synergistic healing effects can be expected as estimated mechanism mentioned above. [24]

4.5 A proposal of the developer of loess bio-ball and its bio-Products

A proposal for special use of loess bio-ball bio-products with some unique effects is introduced briefly the healing effects of each bio-ball bio-products. It can be said that prevention effects of lifestyle-related disease and of recurrence after surgery are unmistakable. The details will be presented at the oral presentation site.

5. Conclusion

The useful characteristics of loess powder for application to environmental friendly bio products can be clarified from measurement data of physico-chemical properties for loess powder.

As an important healing mechanism of far infrared rays by comparing experienced case studies received from many users and loose finding outcomes on loess ball bio products, the following three mechanisms are considered:

1) Improve circulation of blood by solving the stoppage of erythrocytes in the blood,
2) Eliminate pain due to relaxation of muscles,
3) Eliminate sleep and stress through stabilization of the nervous system by early release of harmful substances.

Various experienced cases are unable to explain sufficiently at a moment. Further detailed research and clinical trials are required and this result brings to utilize loess powder more widely as a raw biomaterial.

Acknowledgments

We sincerely express our gratitude to Makino President Mr. Katsunori Makino and Dr. Masakake Kamitani who gave us valuable advice through measurement and discussion of loess powder. We want strongly to make joint research work with you in any type of academic or business.

References

1. http://www.kfir.or.kr: "Information of Far-infrared rays and Anion" (2012).
2. http://www.enseki.or.jp: "Basic of Far-infrared rays" (2012).
3. http://www.nasa.gov/mission pages/herschel/multimedia/pia15260b.htmlNASA
4. Dr. Habil. Christiane Wetzel, "Study of the effectiveness of bio-resonance therapy", pp. 106~107, Fraunhofer FEP, Annual Report (2011).
5. Simon, W.: "New wave healthy method translated by Michio Norose", pp. 12-37, Gendaishorin (2007)
6. http://www.nukuemon.com/ensekigai。html: "Far-infrared rays?" (2012)
8. Woo Sik Choi, Yeon Jin Choi, U-Hyun Paek, Shigeki Toyama, "Characteristics on Honeycomb Structure-like Loess (Hwangto) from Korea and its Applications to Bio-products", The Proceedings of the 50th Symposium on Powder Science & Technology, pp. 79-83, Kyoto, Japan (2012).
9. Woo Sik Choi, Yeon Jin Choi, U-hyun Paek, Shigeki Toyama: "Characteristics of Honeycomb Structure-like Loess mined in Korea and its Applications to Bio-products", J. Soc. Powder Technol., Japan, 50(9), 682-685 (2013).
10. Woo Sik Choi, Shigeki Toyama, Yeon Jin Choi, Byung Sik Woo, "Infrared rays Radiation Characteristics of Korean Loess and Its Ball and The Effect on Human-Healing", The Proceedings of the 51th Symposium on Powder Science & Technology, pp. 42-46, Shimane, Japan (2013).
11. Woo Sik Choi, Shigeki Toyama, Yeon Jin Choi, Byung Sik Woo, "Water Quality Characteristics of Leachate Water from Yellow Loess and Its Application to Daily Water", The Proceedings of the 52nd Symposium on Powder Science & Technology, pp. 24-28, Himeji, Japan (2014).
12. Woo Sik Choi, Shigeki Toyama, Yeon Jin Choi, "Preclinical efficacy examination on healing practices and experiences of users for pillows and mattresses of loess ball bio-products", Procedia Engineering, 102, 399–409 (2015).
13. Choi, Y.J.: "Opening a world that is afraid of cancer", Power Korea, 4, 69-73 (2012), "Customers details on case experiences", p. 39-49 (2016).
14. Basic Properties data collection of loess measured by Busan center of KBSI (2012)
15. Report on Basic Properties data of loess measured by Makino Inc., Feb. (2016).
16. Certificate of test analysis for loess ball, KICM (2003).
17. http://medi-core.co.kr/kor/thermography.php, "Introduction of IRIS-XP"
18. Y.K. Kim, "Perceived Effects after Using the Loess-Room", YOSONG KEONGANG, 2(2), 25~50 (2001)
19. Fanger O. Thermal Comfort. Copenhagen (1970)
20. Y. Kurazumu and T. Tsuchikawa, Proress in studies on specific data related to heat balance equation of the human body, Jpn., J. Biometor, 41(3), 109-128 (2004)
21. Huh, J.: "Donguibogam translated by Hun Young Cho", pp. 2753-2758, Publisher Yeogang (2005)
22. Fatma Vatansever and Michael R. Hamblin, Review: Far infrared radiation (FIR): Its biological effects and Medical application, Photon Lasers Med; 1(4): 255-266 (2012).
23. Kikuji Yamashita, The Effect of the Far-Infrared rays Energy Radiation on Living Body, InTech (2012).
24. Park, YH: Korea Patent No. 10-1178163, "Manufacturing method of loess ball for bed" (2012).

508619 Mesoporous Silica Particles Used for Amorphous Pharmaceutical Formulations

Abstract:

In this work, we explore a complex strategy for utilization of well-defined mesoporous silica particles for poorly soluble drug amorphization. The strategy includes, from start to finish, the scale-up of the particles production, the drug loading methods and their scale-up, the loaded particles characterization, the final formulation production and its evaluation.

More and more newly discovered APIs are on one hand very potent, but on the other very poorly soluble. Amorphization of the API in the formulation presents a relatively easy route for increasing the apparent API solubility and thus higher bioavailability in the gastro-intestinal tract. Various methods are being utilized in order to prepare the API in an amorphous form and also to stabilize it. However, most of them either have poor efficiency, high cost of manufacture, need to be tailor-made for each API or suffer from recrystallization upon storage. In comparison, mesoporous silica particles present a very good opportunity to be used as an almost universal, easy to produce and easy to deploy carrier of amorphous APIs.

Due to their very good sorption capabilities, mesoporous silica micro- and nano-particles are often being studied for their possible applications in drug delivery. Most of these studies are focused on possible utilization in the intravenous applications. Only a minority of the works focus on the use of silica particles in the oral delivery area, which poses significantly lower amount of challenges before final deployment, than the complex intravenous delivery route. Since the APIs cannot recrystallize once loaded inside the silica mesopores due to the spatial restrictions, amorphized API present in silica doesnâ€™t need to be further stabilized to prevent recrystallization. Furthermore, the high specific surface area of silica particles and high hygroscopicity of silica particles further promote dissolution of the amorphized payload.

In this work, three different strategies of API loading into the silica particles are explored. The first method relies on impregnation of the silica powder dispersed in saturated solution of the API, relying on sorption equilibrium to achieve loading, or using incipient wetness impregnation, where exact amount of highly concentrated API dissolved in a volatile solvent is injected into the silica powder and then evaporated. In the second method, already prepared placebo tablets containing high amount of empty silica particles are impregnated with a solution of API using ink-jet technology. This involves very precise dosing, which provides the opportunity to tailor the final API dosage for each tablet individually. The presence of free pore volume of the silica particles provides the space needed for the API deposition inside the tablet and also ensures the API amorphization. Also, the fact that the API is dosed after the tablet preparation allows for preparation of the desired pharmaceuticals on-demand. This way, only cheap to produce placebo tablets and pure API need to be stored. Lastly, the third method relies also on pre-prepared tablets, only this time containing crystalline API already in them together with empty silica particles. The API is then amorphized from its crystalline state using microwave heating. The API inside the tablet melts and is absorbed by the silica, where it stays in amorphous form.

Three different types of structured mesoporous silica particles are used throughout the experiments. They were chosen with regards to their suitability for the specific application, the simplicity of their preparation and the possibility to scale-up their production. The scale-up strategies of the particle production were investigated in order to ensure high enough production of experimental material and also to explore the future industrialization possibilities. Several experiments were conducted in order to identify the key scale-up parameters followed by advanced computational modelling of the hydrodynamic conditions in the stirred vessels from small laboratory scale to large reactors. The results were used to identify the conditions needed to achieve successful results on large scales. The API loading scale-up possibilities were also proposed and evaluated.

Model tablet formulations for each of the mentioned methods were developed in relation to be able to accommodate high silica content and provide fast disintegration times in order to further speed up the dissolution process of poorly soluble APIs. The prepared particles, the particles loaded with API and the final formulations were characterized in detail and their performance was evaluated, providing more than satisfactory results.

508457 Use of Supercritical-CO_2 Assisted Spray Drying Technology for the Production of Pharmaceutical Powders

Abstract:

Purpose

The main goal of this work is to explore an alternative particle engineering (PE) technology for the production of pharmaceutical compounds with enhanced properties and improved performance while maintaining a high process throughput and yield. In this work, the Supercritical-CO_2 assisted spray drying (SASD) particle engineering technology was used to produce composite inhalation powders. For that purpose, a systematic Quality by Design (QbD) approach using the design of experiments (DoE) tool, followed by a statistical analysis to predict the powder *in-vitro* aerodynamic performance were implemented.

Introduction

In what regards to the inhalation field, the delivery of drugs to the lungs is increasing, either to treat local diseases such as asthma, cystic fibrosis, lung cancer, amongst others, or for systemic drug delivery, such as insulin for diabetes. Drug delivery to the lungs presents several advantages such as the rapid onset of drug action and improved bioavailability by avoiding first pass metabolism, which reduces the total drug load required and minimizes the potential for adverse side effects, leading to a more efficient therapy for the patients.

The new strategies of carrier-free inhaled drug delivery are possible due to advances in the particle engineering field, in which Spray-drying (SD) is one of the most well-known and enabling technologies. However, due to the environmental deterioration, the exploration and use of alternative greener technologies should be encouraged. For this reason, the production of composite particles using supercritical fluids, namely, Supercritical-CO_2 Assisted Spray Drying (SASD) was investigated, using supercritical carbon dioxide ($scCO_2$) as a cosolvent to assist the atomization process, minimizing the use of solvents. $scCO_2$ was selected due to its mild critical conditions (31.4 ºC, 7.4 MPa): it is inert, safe, non-toxic, recyclable, inexpensive and environmental friendly. The use of $scCO_2$ as a cosolvent minimizes the use of organic solvents, decreases the drying temperature required for the solvents evaporation, being a more eco and energy-friendly approach, does not required additional gas for the atomization step, which reduces the consumption of gases and potentially reduces the shear stress induced during the atomization step that may cause protein denaturation. Furthermore, the possibility of using lower drying temperatures is particularly suitable for processing labile drugs such as biologics (e.g. proteins). Additionally, it is a highly versatile technology, able to work with water or organic solvents, enabling the handling of both hydrophobic and hydrophilic drugs, while generating powders with enhanced aerosolization properties.

In the SASD PE technology, the feed solution composed of the solute and solvent or solvents mixture, and the $scCO_2$ feed are independently pumped and mixed inside a heated static mixer to promote saturation of the mixture with $scCO_2$, by adding $scCO_2$ in excess. The resulting mixture is atomized inside the drying chamber through a pressure nozzle. The formed droplets are then dried by a hot drying gas stream to evaporate the solvents, forming dried particles. The drying gas used was air. The particles were then collected by a high efficiency cyclone. The main difference between the SASD and SD technology is the atomization step. In the SD technology the atomization step has only one phase: the liquid disperses and forms droplets that are then dried due to the solvent evaporation, forming particles. In the SASD technology, the liquid is dispersed forming the primary droplets; then the $scCO_2$ inside the primary droplets expands and forms smaller secondary droplets that will form particles upon solvent evaporation.

The main goal of this work was to optimize the SASD technology process parameters for the production of composite particles with enhanced properties and *in-vitro* aerodynamic performance using the DoE tool under QbD principles. A formulation composed of trehalose and leucine dissolved in a water/ethanol system was used.

Experimental methods

Composite particles of trehalose/leucine were produced by a custom-made SASD apparatus. The feed solution was composed of 2% w/w of trehalose/leucine (C_solids) in an 80:20 weight ratio dissolved in 20% w/w ethanol-in-water.

The DoE performed was a full factorial design (8 points plus 2 central points to assess the reproducibility) and the process parameters investigated were: the static mixer pressure (P_sat: 9.0 – 11.7 MPa), the drying gas inlet temperature (T_in: 85 – 150 ºC) and the liquid feed flowrate (F_feed: 3.5 – 7.5 mL/min). The main goal was to determine the most critical parameters.

The powder properties were characterized in terms of particle morphology by SEM, particle size by laser diffraction, bulk density using a graduated beaker, residual water and ethanol content using Karl-Fischer and gas chromatography respectively, mDSC and XRPD for the solid-state characterization and the *in-vitro* aerodynamic performance was evaluated using an 8-stage ACI actuated by a Plastiape device at 60 L/min. The powder fine particle fraction (FPF) was measured below 5 µm and relative to the capsule emitted dose ($FPF_{ED(<5\mu m)}$).

The statistical analysis used to quantify the impact of the input parameters (P_sat, T_in, F_feed) on the output parameter (FPF) was performed using SIMCA v13.0.3.0 software from Umetrics.

Results & Discussion

Process yields as high as 70% (batch size of 11 grams) were obtained. Regarding the powders PS, it was observed that the powders produced at a higher T_in and F_feed presented a larger PS than the powders produced at lower temperatures and lower F_Feed. The higher temperature possibly caused particle inflation without breakage, leading to slightly larger particles while the lower F_feed possibly produced smaller droplets because the ratio F_CO_2/F_feed increased, possibly leading to a stronger secondary atomization that formed smaller droplets and then particles. The particle size span of the SASD particles is smaller than the particles produced by SD using a conventional two-fluid nozzle.

Regarding the residual water and ethanol solvent content, it was observed that the higher the drying temperatures (T_in and T_out) and the lower the F_Feed, the lower the residual water and ethanol content in the final powder. In addition, all powders presented a residual EtOH content below the ICH limits (< 5000 ppm). Globally, it is possible to observe that as expected, the powders BD is inversely proportional with the powder T_in. The powders produced at the higher drying temperatures and lower F_feed values presented a lower BD since higher temperatures promote the production of hollow particles which in combination with a low F_feed, leads to the production of particles with lower water and EtOH content, which further contributes to a lower BD. The same trends were observed in previous work conducted using the Spray drying technology where higher T_in and lower F_feed produces smaller particles.

According to the XRPD and DSC data, and similarly to what was observed when spray drying this formulation, amorphous trehalose and crystalline leucine were observed in all powders.

According to the SEM micrographs all powders seem similar, uniform and spherical with the powders spray dried at higher T_in presenting a slightly smoother surface, in agreement to what would be expected and was observed with conventional SD.

The powders were then characterized in terms of aerodynamic performance where $FPF_{ED(< 5\mu m)}$ values as high as 86% were obtained. The mass median aerodynamic diameter (MMAD) values were also lower than 2.5 µm for all cases. All powders presented a high FPF (76 – 86%), which makes it harder to draw conclusions on the process parameters impact due to the very small variations. Therefore, it would not be possible to make a clear conclusion without applying statistical analysis tools.

The statistical models obtained predicted well the powder properties as well as the powder FPF. The reproducibility trials presented similar FPF values which shows the good process reproducibility. An acceptable model was obtained, being able to predict the powder FPF based on a correlation with the aerodynamic particle size (aPS) – determined based on the powder PS, density and shape factor – which presented a R^2 and Q^2 of 0.658 and 0.545, respectively. It was observed that the main descriptors that influence aPS and therefore the FPF are the F_feed and T_in where it was observed that the higher T_in and the lower F_Feed the higher the FPF. According to the model, the P_sat did not had a significant impact on the powder aPS and consequently on the FPF.

Conclusions

In the proof-of-concept, small and uniform particles (Dv50~1 µm) were produced which resulted in inhalation powders with FPF values as high as 86% by decreasing the F_feed and increasing the T_in. When comparing to powders produced using the SD technology under analogous conditions, similar solid state properties were observed. The SCF technology presents unique features and several advantages particularly relevant for the production of micro and nano particles for manufacture of small molecules and biopharmaceutical compounds. Further studies, outside the analyzed ranges and concerning other process/formulation parameters should be undertaken to better understand this technology and widen its potential to produce particles suitable for other delivery routes such as oral, injectable, nasal and topical amongst others.

508483 Spray Drying Scale-up Based on a Development By Design Approach

Abstract:

The traditional scale-up approach of pharmaceutical manufacturing processes involves substantial testing at scale which leads to the expenditure of tens and sometimes hundreds of kilograms of product in a very expensive and time consuming endeavour. The main reason for this experimental based approach is the underlying complexity of pharmaceutical manufacturing processes, the lack of process understanding and scale independent correlations.

At present, the scale-up paradigm as shifted from this experimental based "Trial & Error" approach to a process understanding QbD (Quality by Design) approach through the advances in PAT (Process Analytical Technology) measurement tools and modelling tools ranging from simple heuristics, to mechanistic and statistical models and to more advanced tools such as CFD (Computational Fluid Dynamics), DEM (Discrete Element Methods) and PBM (Population Balance Models).

Spray drying is one such example of a pharmaceutical manufacturing process where the QbD initiative was instrumental in process understanding and in building mathematical models. Due to its remarkable capability in manipulating powder/particle characteristics such as size, morphology, density and level of residual solvents, it has become one of the most sought after unit operations for the manufacturing of ASDs (Amorphous Solid Dispersions) in order to overcome poor water solubility/bioavailability of oral drugs [1]. Spray dryers in the pharmaceutical industry are available in a wide range of scales: from lab units where milligrams of material can be produced to very large commercial units capable of handling several tons per day. One particular concern during development of pharmaceutical spray dried materials is the effect of scale-up on critical quality attributes, where a less careful scale-up strategy may lead to considerable losses of very expensive materials and jeopardize the timelines of a clinical program. As such, the need for robust process development methodologies is critical to minimize the consumption of materials and reduce development time and time to market. The quantities of API required in the early stages of development are typically small, but may increase by many orders of magnitude as a drug candidate advances through the clinical phases and reaches the market. This requires multiple scale-up steps along the process [2].

With 73% of the novel drugs approved by FDA in 2016 falling into one or more expedited approval categories (Fast Track, Breakthrough Therapy, Accelerated Approval and Priority Review) CMC development is increasingly on the critical path of drug development. Under these programs, work that traditionally takes ten years often must be compressed into two or three, where specialized experience and expertise is required to successfully develop robust and cost-effective processes that assure the same levels of quality [3].

The work presented here describes a development methodology case study for a spray-dried product from the lab to commercialization. This methodology, termed DbD (Development by Design), aims at a process development in a timely and cost-effectively manner with minimal expenditure of time and drug substance during development and encompasses three stages: i) Familiarization Stage at the Lab-Scale: In this stage the goal is to develop scale-independent dimensionless correlations to support the multiple scale-up steps; ii) Scale-up for clinical supply: In this stage the scale-up is performed, through simulation tools, under GMP without experimentation at industrial scale and iii) Process Intensification and Design Space definition: This stage is performed close to validation when the risk for the clinical program is minimal. Experiments are conducted to improve process efficiency and support the definition of the design space. Rather than tens to hundreds of kg, as little as 30 g may be required for testing. These results clearly demonstrate that we have been able to combine thermodynamic, drying kinetics, atomization and CFD (Computational Fluid Dynamics) modelling, extensive process knowledge, and a database of over 1000 successful spray-drying batches to accelerate the development and commercialization of drugs.

[1] Newman, Ann, Gregory Knipp, and George Zografi. "Assessing the Performance of Amorphous Solid Dispersions", Journal of Pharmaceutical Sciences 101.4 (2012): 1355-1377.

[2] Marco Gil, João Vicente, Filipe Gaspar. "Scale-up methodology for pharmaceutical spray drying", Chemistry Today, 28, (2010):18-22.

[3] Food and Drug Administration (FDA), (2016) Novel Drugs Summary.

506750 Large Scale Discrete Element Modeling (DEM) Using Rocky Multi-GPU Solver for Various Process Equipment

Abstract:

Chemical and Pharmaceutical industries have large manufacturing unit operations that involve complex physical processes that are not well understood. For example, there are different particle-level phenomenon at play simultaneously in processes such as granulation or tablet mixing/coating. Scale-up can thus be quite challenging during process development and is often done using large scale design of experiments, which can be difficult and expensive to accomplish. Predictive first-principle models such as the Discrete Element Method (DEM) not only provides insights into the manufacturing process but also reduce the number of trial-error experiments by predicting material behavior in these unit operations.

The DEM is the modeling of particle-level interactions on the computer, but in the past, this method was limited by hardware memory limitations and large simulation run times. In addition, particle shapes were also limited to be either single spheres or a cluster of spheres rigidly bonded together (â€˜gluedâ€™ sphere) as the contact detection algorithms were easier for sphere-sphere interactions. These limitations were overcome, for the first time, in the Rocky DEM software, which uses state of-the-art polyhedral shapes and distributed GPU solver technology. This presentation focuses on highlighting Rocky DEMâ€™s unique shape and multi-GPU features to predict flows in processes such as a commercial scale tablet coater, high shear wet granulator, material crushers, and more.

Rocky DEM was used to model realistic tablet shapes using its polyhedral shape representation abilities in the simulation of a commercial scale (BFC-400) tablet coater (359 kg batch containing 250,000 tablets with 222 vertices each). In addition, fast multi-GPU solvers allowed for the prediction of tablet movements in the coater in reasonable compute times (approx. 3 days using three NVIDIA Tesla P100 GPU cards: 86x faster than an 8-core CPU).

High-shear wet granulation equipment, which is used in the manufacture of solids oral-dosage process to improve the bulk density and flow characteristics of a formulation, was also modeled in Rocky. The challenge of this work was to establish correct scale-up rules by achieving a process level understanding. Granulators at various scales (1 L, 10 L and 150 L) using both dry and wet placebo were simulated and the predictions of particle velocities, stress profiles and residence time distributions were compared with experimental data. The ability to combine the power of several GPU cards allowed the processing of large scale particle count simulations using particle sizes and distributions closer to reality, especially at the 150-L scale, where over 10 Million particles were used. Extensive post-processing analyses were also performed in order to enable direct comparison of particle volume fractions and velocity profiles. In addition, prediction of the stress profiles and collisional behaviors, which are impossible to obtain experimentally, increased process understanding and reduced the number of experiments during scale up studies.

507045 Scale up of Dry Granulation from Pilot Scale to Commercial Manufacturing

Abstract:

Dry granulation (roller compaction) is a common processing technique used in the pharmaceutical industry to improve the flow of moisture sensitive drug substances. There are a number of manufacturers of this equipment. While the basic principle of roller compaction is the same for most equipment each device has its own unique operating attributes which differentiates it from another. These differences can present challenges when scaling up during development and technology transfer.

A case study is presented outlining the techniques used to transfer a roller compaction process from an Alexanderwerk WP120 in the development labs to a WP200 in a commercial manufacturing environment. Additionally techniques will be discussed which look at transferring from an Alexanderwerk to a Gerteis unit and subsequently scaling from a Gerteis Mini-Pactor to a Macro-Pactor. Ribbon density and particle size distribution of the granulation along with dissolution of the tablets produced were the primary outputs used to determine whether material manufactured on any unit was equivalent and/or appropriate.

506133 Role of the Ejection Process in the Capping of Pharmaceutical Biconvex Tablet after Compaction: A FEM Study

Abstract:

Tablet is the most popular pharmaceutical form. Its manufacturing using die compaction has been used for more than a century. Nevertheless, classical problems encountered during compaction are still to be solved. Among them, capping corresponds to the separation of the top layer of the tablet and can be observed just at the ejection from the die or some days after. A large number of articles have been written on the subject to establish correlations between powder/process parameters and capping tendency of the tablet. Nevertheless the problem is still not fully understood.

One of the known parameter influencing capping is the shape of the tablet. It is well known that biconvex tablets are more prone to capping than flat-faced ones. The reason is the mechanism that makes the tablet break. Various publications have described this mechanism. Forty years ago, Hiestand et al. explained that during relaxation, the material in the cup could expand radially whereas the main body could not explain radially. This promotes the development of large shear stresses that can lead to capping. More recently, FEM modelling was used to illustrate this mechanism. It confirms that the elastic recovery of the tablet during unloading promotes the development of a high shear stress at the limit between the land and the cup of the tablet. This shear stress is likely to be responsible of the capping of biconvex tablets which can thus be considered as a failure in shear.

Considering this mechanism, if the compaction is applied symmetrically, the shear stress considered is the same on both side of the tablet. Capping should thus appear on both side of the tablet. Nevertheless it is well known that, sometimes, capping can be asymmetric, i.e. failure occurs only on one side of the tablet. When this phenomenon happens, capping always occurs on the side that is ejected first (i.e. the â€œupper sideâ€ on most of the machines). This dissymmetry is not explained by the mechanism proposed above.

In this presentation we will study how the ejection part of the compaction cycle could be responsible of this asymmetrical failure. In fact the ejection process is clearly asymmetric. It is thus interesting to study how the stresses inside the tablet vary during the ejection process, especially the shear stress at the limit between the land and the cup of the tablet.

Experimental characterization of the stresses inside the tablet during the process is difficult. But, as already demonstrated in a lot of studies, numerical modelling using finite element method (FEM) makes it possible to access the stresses inside the tablet during the whole compaction cycle. In this work, FEM modelling, based on the Drucker Prager Cap model for the powder behavior, was used to study the influence of the ejection process on the stresses inside the tablet and to explain the role of the ejection process in the capping of biconvex tablet. Afterwards, influence of some process/product parameters on the capping tendency of biconvex tablet were discussed, based on the phenomena found in the simulations.

In a previous publication, we focused on the elastic recovery of a biconvex tablet during decompression using FEM. One of the findings was the fact that the contact between the punch (upper and lower punches) and the tablet was not lost at the same time on all the points of the surface. The contact was lost gradually from the land to the center of the cup, due to the elastic recovery of the cup of the tablet. A logical consequence of this typical elastic recovery is that the radius of curvature of the tablet at the end of the decompression is lower than the radius of curvature of the punch.

This elastic recovery has also consequences on the first part of the ejection process. During this part the lower punch moves up in the die. As the curvature of the tablet is higher than the curvature of the punch (i.e. the radius of curvature is smaller), the contact between the punch and the tablet will begin at the center of the cup. Because of the frictions between the tablet and the die and because of the residual die wall pressure (RDP), a certain force (ejection force) is needed to move the tablet upward. Until this force is reached, the band of the tablet will not move even if the punch is moving up. When the lower punch moves up, the contact area between the punch and the tablet increases without any movement of the tabletâ €™s band. The consequence is a deformation of the tablet during this part.

Thanks to FEM modelling, it is possible to study the influence of this deformation on the stress distribution inside the tablet. Attention was focused on the shear stress at the limit between the land and the cup on the upper and lower sides of the tablet.

During the beginning of the ejection, it can be seen that the surface of the cup is moving whereas the land is not. This represents the deformation of the tablet. The natural consequence of this deformation is an increase of the shear stress on the upper side of the tablet. At the same time, the reverse trend is observed on the lower side of the tablet, i.e. the shear stress decreases.

At some point, the force needed to move the tablet is reached, and the tablet begin to move upward. During this movement, the value of the shear stress remains nearly constant on both sides of the tablet.

This FEM simulation of the first part of the ejection process makes it thus possible to understand the influence of the ejection on capping of biconvex tablets. During the first part of the unloading, the shear stress, that is responsible of the failure of the compact, increases on the upper side of the tablet and decreases on its lower side. The shear strength of the tablet may thus be reached because of the deformation of the tablet. This would promote capping only on the upper side of the tablet as sometimes observed during manufacturing.

This mechanism makes it also possible to understand the influence of various process/product parameters on the capping tendency. As explained above, the deformation of the tablet during the beginning of the ejection explains that capping occurs on the upper side of the tablet. The deformation of the tablet is due the force applied by the lower punch. It is obvious that a higher force applied by the lower punch promotes a greater deformation of the tablet. So, all product/process parameter that would increase the ejection force would favor capping.

First, ejection force obviously increases along with the friction coefficient between the tablet and the die. A high friction coefficient is likely to promote capping. This result indicates that a bad lubrication, which will increase the friction coefficient, increases the capping tendency of biconvex tablets. This trends is already well known and the mechanism presented in this presentation explains it.

The second point, which is of great importance, is the influence of speed. It is well known that increasing the speed favors capping occurrence. In fact, increasing the compression speed increases the ejection force. Some authors explained the increase of the ejection force along with the speed through problems of lubricant migration during the compression. If this explanation can be imagined there is, to our knowledge, no experimental data to prove this hypothesis. Another explanation for this phenomenon can be proposed. This increase could be simply due to an inertia effect. In fact due to the high speed of the punch, there is an overshoot due to the fact that the force is not transmitted instantaneously. This fact can be observed experimentally.

When slow speed is used, the ejection signal is similar to the one obtained in the simulation. The force increases and then reaches a plateau as predicted by the coulomb friction theory. On the contrary, the ejection signal obtained at high speed has a totally different shape. First of all, the peak force is much higher than in the case of a low speed. Moreover, after the peak, the force oscillates. This pattern is typical of a stick-slip phenomenon. This signal means that during ejection, the tablet is subjected to an oscillating deformation that will induce successive increase and decrease of the shear stress on the top of the tablet. This kind of mechanical solicitation is of course not in favor of the tablet integrity and is likely to increase the capping tendency.

Particle Measurement in High-Temperature Gas based on Light Scattering

Li-feng LU[1,3], Xiao-lin WU [1,3], Zhong-li JI [1,2], Ming-xing WANG[1,3]

[1] Department of Chemical Engineering, China University of Petroleum, Beijing, P. R. China

[2] Department of Mechanical and Transportation Engineering, China University of Petroleum, Beijing, P. R. China

[3] Beijing Key Laboratory of Process Fluid Filtration and Separation, China University of Petroleum, Beijing, P. R. China

Abstract: In order to achieve the particle measurement at 650 °C, a single particle measurement method and device suitable for high-temperature gas were developed. The influence of gas temperature on optical sensor of optical particle counter (OPC) was investigated in detail based on geometric optics theory and heat transfer theory. The thermal expansion of aerosol tube along the airflow direction was caused by the heat conduction between the tube and the gas, resulting in the position of the light apertures to shift to obscure the partial light beam. The invar material can be adopted to eliminate the above mentioned disadvantages. According to the principle of gas state equation, the optical parameters (refractive index) of the gas at different temperatures change with the change of its state parameters. Based on the geometrical optics theory and the geometric mathematical relationships, the theoretical and experimental dynamic models of the optical measuring volume with the change of the gas temperature were established. The established models were then used to analyze the influence of gas temperature on the optical measuring volume. The results showed that the influence of the temperature on the optical system is negligible. Meanwhile, particle detection experiments were conducted at different temperature to achieve a good performance of OPC.

Keywords: High-temperature gas; Optical measuring volume; Dynamic model

1 Introduction

With the continuous improvement of environmental protection and the safety requirements for industrial production, the technology of high-temperature purification is drawing more and more attention, especially in the fields of clean coal power generation, coal gasification, petrochemical industry and metallurgy [1]. For example, nuclear power plants often produce a large number of low radioactive solid wastes, and the combustible wastes usually account for a considerable proportion. The incineration and pyrolysis are effective treatment techniques for the combustible waste, during which the temperature of flue gas is 600~800 °C. With the aim to purify the flue gas, the German Karlsruhe Nuclear Research Institute, the French nuclear power station and the Japanese nuclear power station have installed the silicon carbide ceramic filter tube, aluminum silicate fiber and alumina ceramic filter tube at gas temperatures above 650 °C, respectively, and investigated their properties properties [2-4]. These filters have a direct influence on the purification quality of gas and the particle concentration in the downstream. In an industrial cracker system, the particle concentration is required to reach 0.014 mg/m^3 after the filters; the Integrated Gasification Combined Cycle (IGCC) requires the particle size to be smaller than 1 um with a particle concentration of less than 2.2 mg/m^3 and in the range of 1~10 um with less than 5.8 mg/m^3, and to remove the dust size of larger than 10 um [5]. In a Shell Coal Gasification Process (SCGP), the particle concentration is required to be less than 20 mg/m^3 [6]. In order to detect the concentration and size of the particles in pipes in real time for calculating the separation efficiency of evaluating the purification performance of the filtration system, the particle detection technology is needed to be improved [7].

The widely-used particle detection devices in industry are mainly based on the principle of light scattering [8-10], and the particle size and concentration are obtained through an inversion of the scattering light intensity and particle size, which are suitable for the conditions of below 200 °C. The method of first cooling and then detecting can be adopted for the particle detection under higher-temperature conditions, but droplet condensation and sublimation will inevitably occur and result in an incorrect particle samples.

With the aim to achieve the particle detection at 650 °C, an optimization was conducted on the existing particle detector (OPC) in this paper. And based on geometric optics theory [11] and heat transfer theory, a detailed research was performed on the influence of gas temperature on OPC optical sensor, including the dynamic change of the aerosol tube and the optical measuring volume (OMV). Finally, the experiments were conducted on the on-line particles measurement of different temperatures.

2 Experimental setup and method

Figure 1 gives the experimental setup for detecting particles in high-temperature gas, mainly including the heating unit, optical detection unit and gas purification and processing unit.

The heating unit consists of the vibration particle feeder and the air heater. The gas brings the particle out of the feeder, and is heated up to 1000 °C by the heater, which is monitored quickly by a PID temperature control instrument with easy operation.

The optical detection unit is used to achieve the experimental measurement of the characteristics of the OPC optical system and the on-line detection of particles, including the optical sensor and the OPC control host. The optical sensor is shown in Figure 2, which is composed of an aerosol tube and optical system. Aiming to avoid the influence of high-temperature gas on optical system, the aerosol tube (shown in Figure 3) has been developed, and is in a distance of 15 mm to the optical lens. In addition, the insulation coating with lower thermal conductivity (-80-1700 °C) is adopted to treat the outer wall of the aerosol tube. A K-type micro thermocouple was bent and installed into the light aperture to monitor the gas temperature at the center of the aerosol tube in real time without affecting the airflow and optical paths. Figure 4 shows the optical system of the sensor, which is composed of an incident light path and an observed scattering light path. Both the paths are arranged in 90° with the same optical lens and elements [12-14], and the intersected area is the center of the aerosol tube to form an OMV [15]. The intersected area is enough small and allows particles to pass through it one by one [16, 17]. The OPC host consists of the light source system (wavelength: 300-700 nm) [18], the photoelectric detection system and the signal analysis and processing system. In the particle detection, particle is illuminated by the incident light when passing through the OMV, and the scattering light is collected and transmitted to the photoelectric detector to achieve the conversion of a pulse voltage signal from an optical signal. Finally, the tested particle size is calculated based on the calibration curve of particle size and voltage amplitude [14, 19].

The gas purification and processing unit is composed of a vacuum pump, a flow controller, a dust collector and a heat exchanger. The high-temperature gas form the aerosol tube is completely cooled during the heat exchanger, the flow rate of which is adjusted by the flow controller. In order to avoid the damage of particles on the pump, the dust is captured and collected by a filter in the dust collector.

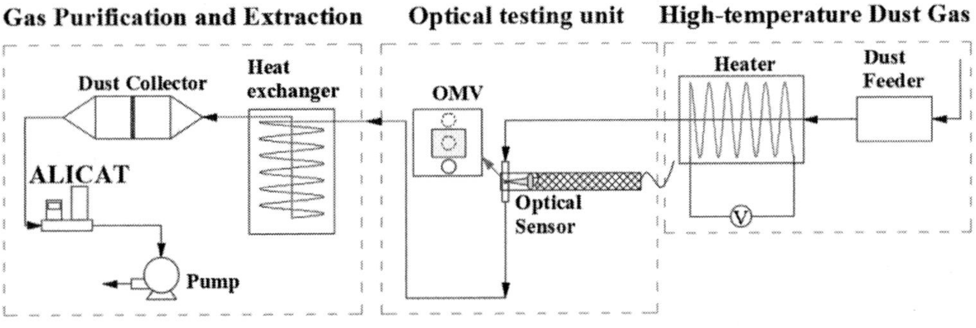

Figure 1. Experimental setup for detecting particles in high-temperature gas

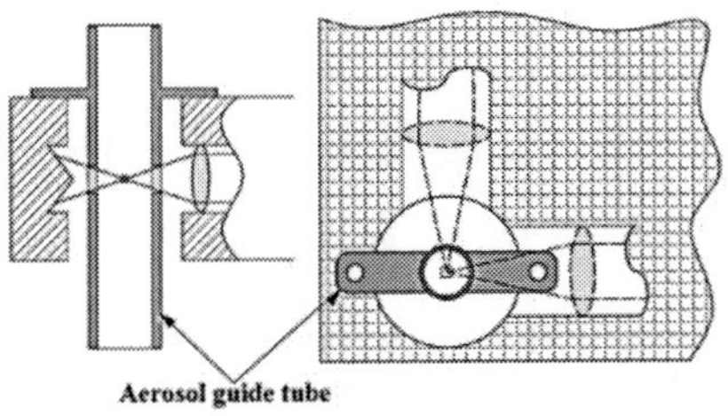

Figure 2. Optical sensor

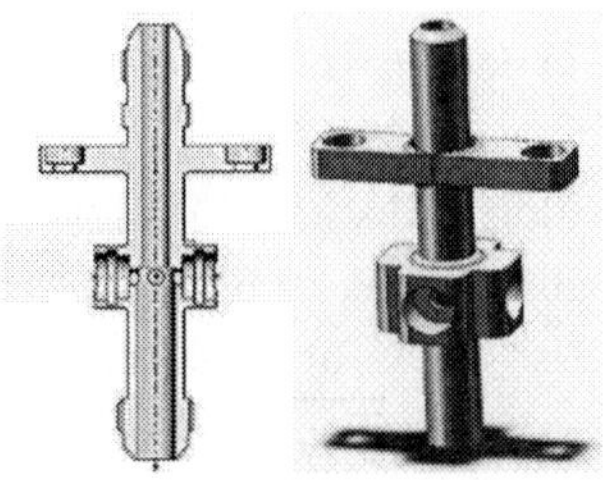

Figure 3. Aerosol tube for high-temperature gas

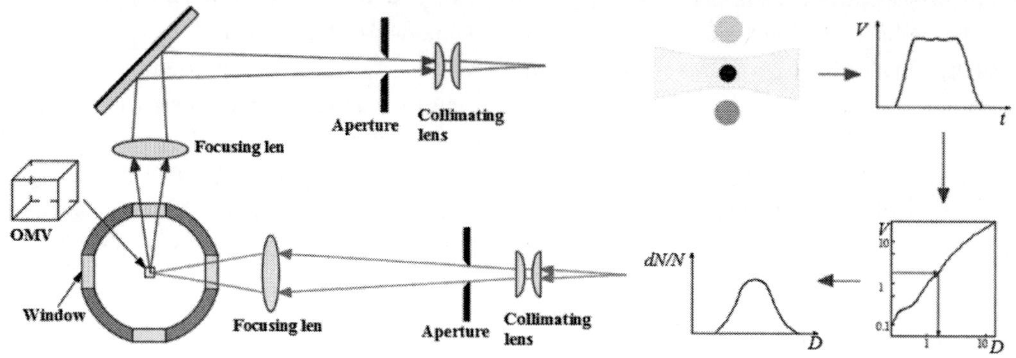

Figure 4. Optical system and principle of OPC

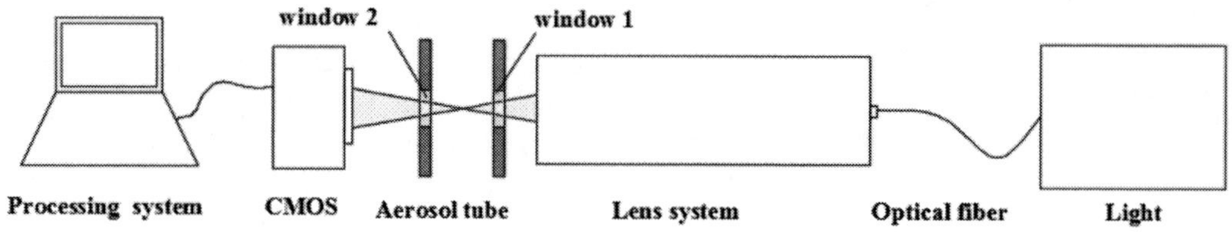

Figure 5. Evaluation and analysis device for light beam

3 Analysis and discussion

3.1 Influence of temperature on aerosol tube

The aerosol tube shown in Figure 3 has four apertures along the center axis, in which the glasses with a thickness of 2 mm were installed, as shown in Figure 4. When the gas flows in the tube, it can reduce the damage of high-temperature gas to the optical system elements. But the heat resistance of the tube has a direct influence on the imaging of the optical system at the center of the aerosol tube.

Figure 5 gives an evaluation and analysis device for light beam, including the light source system, the light path system, the aerosol tube and the light beam analyzer (CINOGY CMOS-1201, Germany), which can be equipped at the optical testing unit as shown Figure 1. For the gases at different temperatures, the light passes through the glasses in the aerosol tube and images on the CCD of CMOS-1201 to detect the spot of the light beam in real time.

After eliminating the background noise, the spot image at different temperatures can be acquired. When the aerosol tube is made of 304 stainless steel, the measurement results are shown in Figure 6. It can be found that the shape of the light spots is distorted and shows the imperfection of the imaging with an increase in gas temperature. That's due to thermal expansion along the airflow direction caused by the change of gas temperature, which results in the shift of light apertures position to obscure the partial light beam. A bad result must be acquired for smaller particle size and lower particle count efficiency because of the decreased incident light. Meanwhile, the shape of the OMV was changed to greatly reduce the performance of OPC detection by a failure of eliminating the error of edge detection in optical system.

With the aim to eliminate the above mentioned disadvantage, the invar material with a linear expansion coefficient of 1.6×10^{-6} m/°C was adopted for high-temperature experiment. The results are shown in Figure 7. At different temperatures, no deformation was found in the shape of light spots, indicating that the material can meet the testing requirement for high temperature up to 650 °C with a good heat resistance.

Table 1. Comparison of the application conditions of the two materials

Material	Light divergence angle /°	Gas temperature /°C
304 (Stainless steel)	<10	<200
4J36 (Invar)	<30	<700

The application conditions of the two materials are

summarized in Table 1. Compared with the stainless steel, the invar material is of great significance for the development of high-temperature detection device.

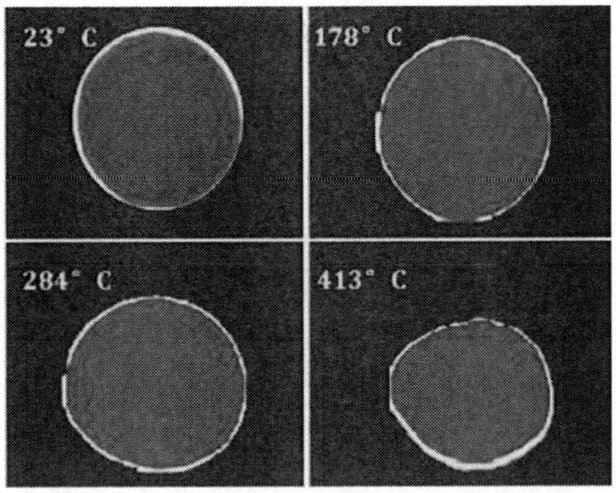

Figure 6. The spot images measured at four different temperatures of gas for 304 stainless steel.

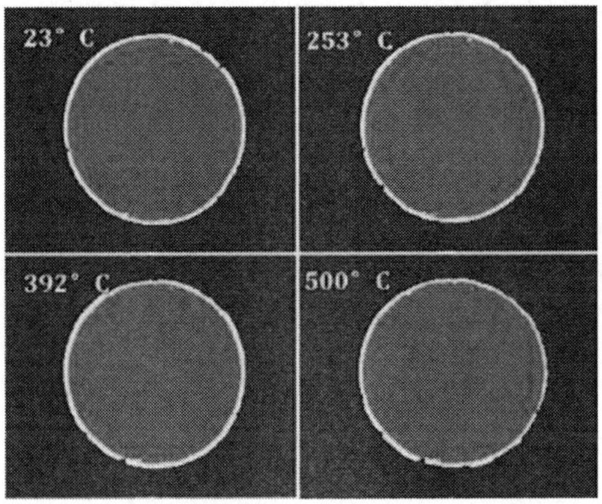

Figure 7. The spot images measured at four different temperature of gas for 4J36 (Invar).

3.2 Influence of temperature on optical system

In general the optical sensing system in the particle detection device is only applied at atmospheric temperature, but can also satisfy the temperature up to 150 °C. Under high-temperature condition, the gas volume is going to expand with the reduction in the gas density. The transmission characteristic of light beam could be changed, resulting from the decrease in the optical performance of the sensor. In order to investigate the influence of temperature on optical system, the specific causes are analyzed by theoretical method, and the extent of its influence is determined through experimental verification.

3.2.1 Theoretical method

By calculating the refractive index of the gas at different temperatures, the change of the OMV with the temperature can be obtained.

a) Calculation of gas parameters

The gas medium is air, so its refractive index can be calculated by the formula proposed by Rueger at the working temperature, given by:

$$n_{ph} = \frac{273.15}{1013.25} \cdot P \cdot N_{sph} \cdot (a_0 + \frac{a_1}{T} + \frac{a_2}{T^2} + \frac{a_3}{T^3} + \frac{a_4}{T^4}) \times 10^{-6} + 1 \quad (1)$$

Where P is the gas pressure; a_0, a_1, a_2, a_3 and a_4 are the correlation coefficient and constant; N_{sph} is the gas refractive index at atmospheric temperature, given by:

$$N_{sph} = (n_{sph} - 1) \times 10^6 = 287.6155 + \frac{1.62887}{\lambda^2} + \frac{0.01360}{\lambda^4} \quad (2)$$

The wavelength range is 400-700 nm. The gas refractive indexes at different temperatures in the range of 25-650 °C are calculated by Eq. 1 and 2. The results are shown in Figure 8. It can be found that the refractive index of air decreases with the increase in the gas temperature. And it can be inferred that the light divergence angle at high temperature is going to increase compared with that at atmospheric temperature. Consequently, the position of the beam focus is offset, leading to larger optical measuring volume, as shown in Figure 9.

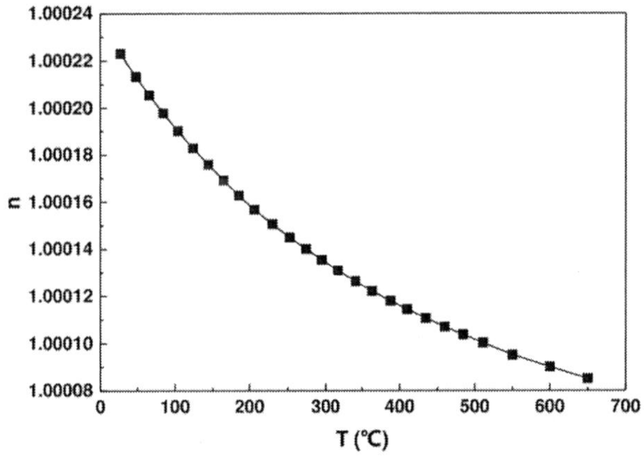

Figure 8. Gas refractive index at different temperature

In order to analyze the above unfavorable influence, a model of the OMV and gas temperature is needed to be built, mainly including the functions of the determination of the optical parameters and the calculation of the dynamic variable.

b) Calculation of optical parameters

On the basis of the Lorentz-Lorenz formula and actual gas state equation, the relationship between gas pressure and refractive index is given by:

$$\frac{n^2-1}{n_0^2-1}=\frac{\rho}{\rho_0}=\frac{PT_0Z_0}{P_0TZ} \quad (3)$$

Where T_0, P_0, ρ_0, Z_0 and n_0 are the gas temperature, pressure, density, compression factor and refractive index under atmospheric condition, respectively; T, P, ρ, Z and n are the gas temperature, pressure, density, compression factor and refractive index under high-temperature condition, respectively.

The light divergence angle is calculated by the refraction law shown below:

$$n_0'\sin\alpha = n\sin\beta \quad (4)$$

Where n_0' and α are the air refractive index and light divergence angle under atmospheric condition; β is the light divergence angle under high-temperature condition.

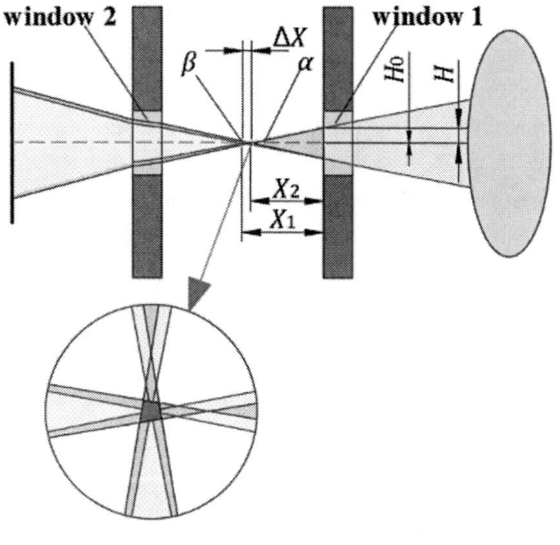

Figure 9. Schematic of the enlargement of the OMV

c) Calculation of dynamic variable

Based on the geometric optical theory and the mathematic relationship, the distance between the light focus position and the glass window 1 is triangular with the equivalent spot radius of the inner side of the glass window 1. According to X_1 and X_2 (the distance between the light focus position and the inner side of the glass window 1 under atmospheric condition and high-temperature condition), the theoretical model of the OMV with gas temperature was built. X_1 and X_2 are calculated by:

$$X1=\frac{H-H_0}{\tan\alpha} \quad (5)$$

$$X2=\frac{H-H_0}{\tan\beta} \quad (6)$$

The change of the light focus position ($\Delta X=X1-X2$) can be calculated by Eq. 5 and Eq. 6.

The change of the equivalent spot diameter of the OMV can be calculated based on the triangle relation, as shown below:

$$\Delta D = 2\Delta X\tan\beta \quad (7)$$

Eq. 6 and Eq. 7 are the formulas of the theoretical dynamic model of the OMV with the change of the gas pressure.

Using the model, the optical parameters at different gas temperatures in the range of 25 ~ 700°C were calculated with three light divergence angles (4.02°, 9° and 13°, tested under atmospheric condition, respectively). The results are presented in Figure 10. It can be found in Figure 10 that ΔX increases with the gas temperature with approximate two patterns. In addition, ΔD changes by a smaller margin (less than 1 um). However, ΔD and ΔX satisfy a trigonometric relation. It can be also noted that the enlargement rate of the OMV volume (δ) has a small increase with the increase in gas temperature. Under the constant temperature, δ is going to increase for the larger light divergence angle, but it is still relatively smaller (<1%) compared with the OMV at atmospheric condition. Consequently, the influence of the gas temperature on the OMV can be ignored based on the above theoretical analysis.

3.2.2 Experimental method

In order to verify the reliability of theoretical analysis, an experimental dynamic model of the OMV with gas temperature variation was established based on the image analysis and the mathematical geometry relations, including the measurement of the change of light spot, the calculation of the geometric relation and the calculation of the relative variable.

By the refraction law, the changes of light spot on the inner and outer side of the glass window 2 under high-temperature condition (ΔD and $\Delta D'$) meet $\Delta D=\Delta D'$. Taking the light focus position as the central axis, a triangle relation was formed by the distances between the light -

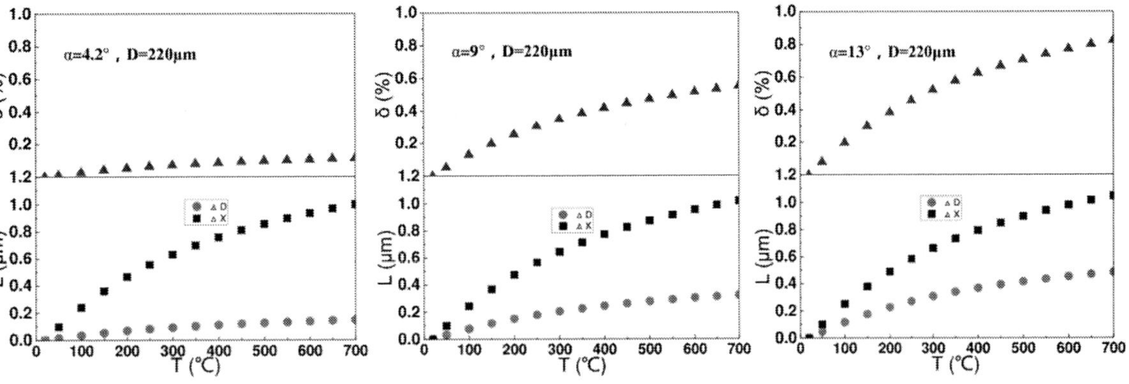

Figure 10. Results calculated based on theoretical dynamic mode with three light divergence angles (4.02°, 9° and 13°)

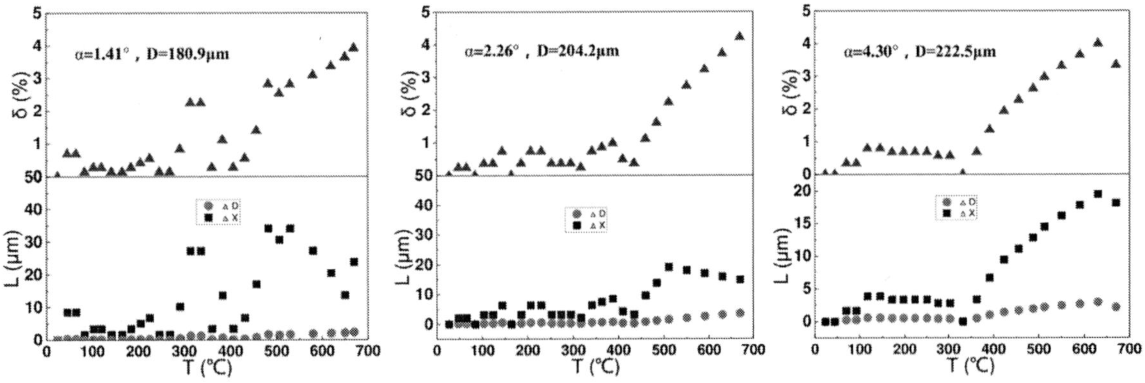

Figure 11. Results calculated based on experimental dynamic model

focus position and the inner side of the glass windows 1 and 2 (x and y), the difference of the light spot equivalent radius on the inner side of the glass windows 1 and 2 (H and h, $h = H - \frac{\Delta D}{2}$) and the equivalent radius of the OMV (H_0). Therefore, we have the following equations.

$$\frac{H - H_0}{h - H_0} = \frac{y}{x} \tag{8}$$

$$x + y = k \tag{9}$$

$$\beta = \arctan\left(\frac{h - H_0}{x}\right) = \arctan\left(\frac{H - H_0}{y}\right) \tag{10}$$

where β is the light divergence angle under high-temperature condition; k is the distance of the inner sides of the glass windows 1 and 2.

The changes of ΔX and ΔD under high-temperature condition to those under atmospheric condition are shown as follows:

$$\Delta X = \frac{y - x}{2}, \text{ or } \Delta X = y - \frac{k}{2}, \text{ or } \Delta X = \frac{k}{2} - x \tag{11}$$

$$\Delta D = 2\Delta X \tan\beta = \Delta D' \tag{12}$$

Eq. 11 to Eq. 12 are the formulas of the experimental dynamic model.

Using the above established experimental dynamic model, the ΔX, ΔD and δ of four cases in Table 2 were calculated, and the results are shown in Figure 11. It can be found that these three parameters increase obviously with the gas temperature increasing, but cannot meet the two function growth mode as the results of theoretical model. That may be caused by the reasons listed below: 1) the fluctuation of gas flow in the pipeline causes the unstable gas temperature field; 2) the beam quality analyzer is slightly influenced by other vibration sources; 3) the aerosol tube is affected by the gas flow fluctuation. But according to the experimental results, the maximum values of ΔX and ΔD are very small (less than 50 μm and 5 μm at the temperature below 700 °C, respectively) compared with the equivalent diameter of the OMV. Consequently, the ratio of the volume of the OMV is less than 5%. To sum up, the gas temperature has greatly weakened influence on the OMV, which can be ignored in actual technical development.

Table 2. Optical parameters of experimental model

number	light divergence angle /°	spot equivalent diameter /μm
1	1.41	180.9
3	1.94	285.1
4	2.26	204.2
5	4.30	222.5

3.2.3 Experiments

The performance of the optical sensor is investigated at different temperatures up to 650°C. The experimental results showed that the temperature of the optical sensor is around 50°C with the gas temperature in the range of 200-300 °C, and is going to increase about 10 °C at the gas temperature of 400-500 °C. The temperature is in the controllable range with a maximum value of 70 °C under the gas temperature of 650 °C. Therefore, the developed optical sensor can be used for the detection of particles in high-temperature conditions for a long time.

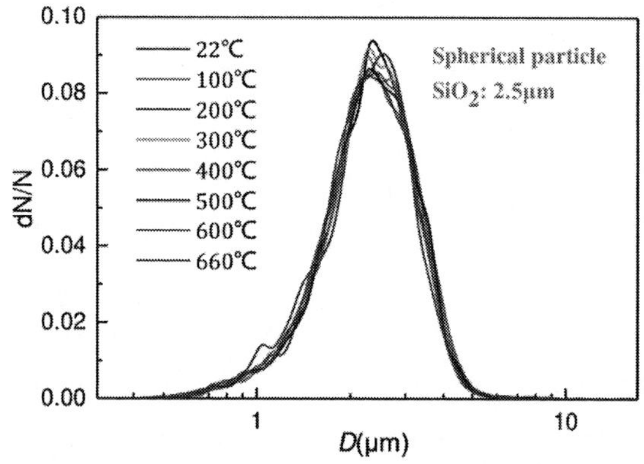

Figure 12. The results obtained under high-temperature condition

In the experiment of particle detection, the standard silica spherical particles with a strong temperature resistance ($D_{particle}$=2.5 μm) were tested at different gas temperatures. The testing time is 60 s. Figure 12 shows the test results. It can be found that all of the particle size distributions are in good agreement and have a good normal distribution, which meets the test results of single particle size, and the peak particle size is about 2.5 μm. Consequently, the optical sensor worked well for the particle detection under higher-temperature condition.

4 Conclusion

The emission of high-temperature flue gas causes serious air pollution and partial waste of available heat. In order to meet the relevant emission standards, additional purification units (high-temperature ceramic filters, etc.) are installed in the industrial process to collect the dusts in the flue gas. As the purification units are important for controlling the dust concentration, it is necessary to calculate the separation efficiency through a real-time measurement of the dust concentration in the flue gas in the upstream and downstream of the purification equipment.

In present work, a single particle measurement method and device suitable for high-temperature condition were developed. In order to complete the particles measurement at 650 °C, the beam images in optical sensing system under different temperature conditions were measured in real time by CMOS sensor. It was found that the beam images were deformed when the light divergence angle increased to more than 10°, and the deformation became more serious as the temperature increased. Depending on the principle of heat transfer, thermal expansion along the airflow direction caused by heat conduction between the duct and the gas is the main cause of this phenomenon. By changing the aerosol tube material t o one with an ultra-small linear expansion coefficient (Invar: 4J36), the optical system can be applied to the condition with temperature up to 700 °C, and in particular the light divergence angle extends to 30°.

According to the principle of gas state equation, the optical parameters (refractive index) of the gas at different temperatures change with the change in its state parameters. According to the results calculated by the formula of Ruegur's atmospheric refractive index, it was found that the refractive index of the gas decreases with the increase in temperature, which leads to the increase in the divergence angle of the beam in the aerosol tube, and shift of the position of the optical beam. Based on the geometrical optics theory and the geometric mathematical relationships, the theoretical dynamic model of the optical measuring volume with the change of the gas temperature was established. By calculating the amount of the beam focus offset and the volume of the optical measuring volume at different temperatures, it was found that the optical measuring volume increases with the temperature increasing. However, when the gas temperature was 650 °C, the increase rate of optical measuring volume was extremely small, so the influence of the temperature on the optical system can be negligible.

The performance of the whole detection system was tested by using 2.5 μm standard spherical silica particles. The satisfying measuring results were obtained and they could be applied to the actual on-site measurement at the temperature of 650 °C.

Acknowledgement

The authors thank the supports of "The National Key Research and Development Program of China (NO.2016YFB0601100)"; and would also like to thank the laboratory and staff of the Beijing Key Laboratory of Process Fluid Filtration and Separation, China University of Petroleum.

Reference

[1] J. Meyer, H. Umhauer, K. Reuter, A. Schiel, G. Kasper, M. Förster, Concentration and size measurements of fly ash particles from the clean gas side of a pressurised pulverised coal combustion test facility, Powder Technology, 180 (2008) 57-63.

[2] H. Leibold, F. Dirks, V. Rüdinger, Particulate emissions from a LLW incinerator and off-gas cleaning with a new type of ceramic candle filter, Waste Management, 9 (1989) 87-94.

[3] B.L. Perkins, Incineration facilities for treatment of radioactive wastes: a review, Incinerators, (1976).

[4] D. Fino, N. Russo, A. G. Saracco, V. Specchia, Multifunctional Filter for Treatment of the Flue Gases from Municipal Waste Incinerators, Industrial & Engineering Chemistry Research, 44 (2005) págs. 9542-9548.

[5] Scheibner, Wolters, Schumacher hot gas filter long-term operating experience in the nuon power buggenum IGCC power plant, (2002).

[6] J.M. Wang, Shell coal gasification technology and its application in China, Coal Quality Technology, (2006).

[7] Q. Xu, Z. Ji, L. Yang, Performance assessment of high temperature flue gas on-line particle size analyzer, Ciesc Journal, 63 (2012) 3506-3512.

[8] B.Y.H. LIU, V.A. MARPLE, K.T. WHITBY, N.J. BARSIC, Size distribution measurement of airborne coal dust by optical particle counters, Aihaj, 35 (1974) 443-451.

[9] R. Maus, F. Löffler, Size analysis of airborne bacteria by an optical particle counter and an impactor, Journal of Aerosol Science, 25 (1994) 141-142.

[10] B. Sachweh, H. Büttner, F. Ebert, Improvement of the resolution and the counting accuracy of an optical particle counter by fast digital signal recording, Journal of Aerosol Science, 20 (1989) 1541-1544.

[11] E.W. Max Born, Principles of Optics, Cambridge University Press, Cambridge, 1999.

[12] S. Hochrainer, Further developments of the particle counter sizer PCS-2000, Journal of Aerosol Science, 31 (2000) 771-772.

[13] B.S. Li, Z.L. Ji, L. Mei, An approach for correcting particle size distributions measured by the optical particle counter welas 2000, Aip Conference, 2010, pp. 686-689.

[14] J. Stier, M. Quinten, Simple refractive index correction for the optical particle counter PCS 2000 by Palas, Journal of Aerosol Science, 29 (1998) 223-225.

[15] H. Umhauer, Particle Size Distribution Analysis by Scattered Light Measurement Using an Optically Defined Measuring Volume, Journal of Aerosol Science, 14 (1983) 765-770.

[16] B. Sachweh, H. Umhauer, F. Ebert, H. Buttner, R. Friehmelt, In situ optical particle counter with improved coincidence error correction for number concentrations up to 107 particles cm-3 quantification and elimination of the effect of particle shape and structure, Journal of Aerosol Science, (1998) 1075-1086.

[17] D.I.J. Raasch, D.I.H. Umhauer, Errors in the Determination of Particle Size Distributions Caused by coincidences in optical particle counters, Particle & Particle Systems Characterization, 1 (1984) 53-58.

[18] H. Buettner, Measurement of the Size of Fine Nonspherical Particles with a Light-Scattering Particle Counter, Aerosol Science & Technology, 12 (1990) 413-421.

[19] D. Holve, S.A. Self, Optical particle sizing for in situ measurements Part 2, Applied Optics, 18 (1979) 1632-1645.

508245 An Experimental Investigation of Single Droplet Drying Above Boiling Point

Abstract:

An experimental method has been developed to enable dried particle structure, properties and drying rate to be predicted based on droplet drying history, particularly focusing on effects driven by boiling including bubble nucleation, growth and collapse. The method involves a single droplet drying rig which will be used to investigate drying kinetics above boiling point and across a broad range of drying conditions, drop size and composition space. The drying rig was improved as compared to previous apparatuses reported in literature, in that it is designed to measure droplet drying rates and morphological changes at high temperatures and under conditions that more closely resemble those found in industrial applications. The aim is to conduct drying experiment for droplets as small as 300μm diameter, in an air stream which temperature may be as high as 350°C.

Drying kinetics and morphological evolution will be determined, particularly focusing on the second drying stage as well as inflation/deflation periods. Initially, the constructed drying rig is checked and evaluated by conducting experiments on droplets of pure water. Afterwards, experiments will be conducted on solutions of different materials such as Hydroxypropyl Methylcellulose (HPMC), sodium sulphate and sucrose. An assessment of morphologies, transformations and drying behaviour under isothermal drying would be completed at a wide space of process condition, initial water content, initial drop size and composition space. The investigation will look to link material properties, particle size, initial moisture content and drying history to morphology. Results will be compared with single droplet drying models reported in literature.

An initial investigation has focused on mapping the morphological drying behaviour of HPMC, sodium sulphate and sucrose. It has shown that the three materials show different behaviour; sucrose showed deflation and inflation cycles, sodium sulphate droplets supersaturated and crystallized while HPMC showed inflation and rupture. Differences in drying behaviour is due to difference in material and transport properties. Therefore, the next stage of this work is focused on conducting some physical properties measurement at different water concentrations and temperature. In addition, different materials will be dried on the rig and the drying behaviour will be determined. The experimental data from this work will be used to develop, a universal morphology regime map which will link material properties, particle size, initial moisture content and drying history to morphology.

508482 Characterising Powder Flow in Dynamic Processes

Abstract:

Flow inconsistency of cohesive powders is a major issue for process industries. It is often difficult to predict and control flow of cohesive powders to enable efficient and reliable process operations particularly for dynamic processes such as mixing and filling. Many quantitative and qualitative flow measurement techniques exist that are used for measuring powder flowability under quasi-static conditions. However, there are few available techniques capable of assessing powder flowability under dynamic conditions. The FT4 Powder Rheometer (Freeman Technology), which drives a rotating blade along the vertical axis through a column of powder contained within a cylindrical vessel, is one of the few available devices that is capable of measuring powder flow in a dynamic state [1]. The flow resistance of the rotating blade as it penetrates the powder bed is characterised by flow energy, however, there is lack of understanding in how to relate this measurement to process design.

This work focuses on determining the relationship between the torque and shear stress in the FT4. The effect of particle size on the external flow profile around the blade has been investigated experimentally for glass beads using a high speed camera. The material investigated shows a decrease in downward specific flow energy with increasing particle size; indicating that the increased shear resistance in the bulk for smaller particle sizes exists under dynamic conditions. Particle Image Velocimetry (PIV) analysis indicates an increase in the volume of particles displaced by the blade with increasing particle size for both upward and downward blade movement. An increase in impeller tip speed leads to a reduction in flow energy for this material, and a reduction in the volume displaced by the blade. The prevailing stresses and strains as well as the volume of particles displaced by the blade are estimated by Discrete Element Method (DEM). DEM allows for the determination of a relationship between the flow energy and/or shear stress with particle properties and blade velocity and direction. Furthermore, Positron Emission Particle Tracking (PEPT) analysis of the FT4 has been carried out to determine the radial and axial displacement of particles with respect to the bed height, and to validate the DEM results.

508671 Impact ofÂ Non-Spherical Projectiles on Granular Media

Abstract:

Recently, several studies involving numerical and experimental methods have focused on the study of impact dynamics in both dry and wet granular media. An example of such instances can be an animalâ€™s foot impacting sand during its locomotion. Understanding the physics behind the interaction between granular media and animal feet can be helpful in optimizing locomotion parameters for bio-inspired robots. However, most of the studies involving impact dynamics considered spherical projectiles under different conditions, while characteristic models should involve more complex shapes. Impacting different geometries with conserved density, volume, and velocity on a granular bed may experience contrasting drag forces upon penetration. This is due to the difference in the surface areas coming in contact with the granular media. In this work, non-spherical geometries of following shapes were considered: cylinder, triangular prism, hexagonal prism and cuboid. These geometries were impacted on a loosely packed non-cohesive dry granular bed with the same impact velocities. The effect of surface area in contact with the granular media was analyzed for impact, with the help of penetration depth measurements. We also tested whether the power law for spherical projectiles is applicable for non-spherical projectile impacts.

508664 Computational and Experimental Shear Cell Study with Rigid Cylindrical Particles

Abstract:

Flow properties of particles are commonly assessed through shear cells. Both the Schulze ring shear tester (SRST) and Jenike shear tester (JST) measure the interparticle stresses which are important to the design of many granular flow processes. This work uses precision-cut rigid cylinders of various aspect ratios and sizes to compare the measured stresses between both testers. It is found that the shearing results for both testers are similar, which is important for future work involving flexible biomass materials.

Previous work has sought to use shear cell results to validate various aspects of discrete element method (DEM) simulations. The SRST is recreated in a DEM simulation with rigid glued-sphere cylindrical particles to assess the contact detection method by comparing the experimental and computation results. Also, a parameterization study is performed to examine the effects of coefficient of restitution, coefficient of friction, and other particle properties on the measured steady-state stress. Overall, this serves as an efficient validation of the contact detection method which will be used to simulate complex particle shapes.

508865 Defluidization Behaviour of Industrial Reactive Powders at High Temperature

Abstract:

Fluidized bed reactors are used in a wide range of industrial chemical processes, and they are particularly useful in high temperature systems thanks to their ability to provide high heat transfer rates and rapid solids mixing which lead to isothermal and more controllable conditions. Their applications spread the environmental, chemical, energy and process industries. For example, they are a key technology in the petroleum industry, catalytic chemicals synthesis processes, combustion and gasification of solid fuels (coal, wastes and biomass), metals productions and many more.

Use of fluidized beds in combustion-type operations may lead to particles sintering and bed agglomeration, which again may result in defluidization, unscheduled process downtime and additional costs. However, the details of the physical mechanisms behind such phenomena are still not entirely understood. Despite the countless researches that have been done so far, such processes still disclose unrevealed features and challenge the researchers worldwide.

The objective of this paper is to assess the effect of high temperature on the flow properties and fluidization behaviour of mixtures of carbon coke and two different titanium ores, namely synthetic and natural rutile. To this end, a simple approach for the early detection of agglomeration in the regimes of incipient fluidization -where the carbon combustion causes defluidization- will be followed. Such a monitoring approach is based on the simultaneous measurements of local temperatures and the pressure drop across the particles bed.

507739 Manufacturing and Characterization of Spherical Blend PBT-PC Particles for Additive Manufacturing

Abstract:

Polymer powder-based additive manufacturing technologies often yield excellent mechanical and functional part qualities. One of these technologies is selective laser sintering (SLS), a process employing micron-sized polymer particles, which are selectively fused by a laser [1]. While hardly any boundaries regarding part design exist, there are quite some restrictions concerning the variety of commercially available SLS materials. For the time being, the most widely used polymeric material for SLS is polyamide 12 (PA12) with a market share of roughly 95% [2]. Hardly any blend materials are available; mostly , the processing of multi-material systems via SLS is achieved by simply mixing different polymer powder materials [3].

In this contribution, a new approach to blend particles is introduced. Via co-grinding and subsequent rounding, manufacturing of spherical particles consisting of semi-crystalline polybutylene terephthalate (PBT) and amorphous polycarbonate (PC) thermoplastic is possible. The polymers are jointly comminuted and mixed in a planetary ball mill and the product particles are rounded in a heated downer reactor [4,5]. The thereby achieved intermixed, but not transesterified, PBT-PC composite particles are characterized regarding size and morphology via electron microscopy. Intermixing of the polymers is investigated via staining techniques and Raman spectroscopy. Thermal properties of the blend particles are analyzed via DSC, also in regard to transesterification in SLS process-like conditions. Furthermore, flowability and size distribution of the blend particles are examined. This way, we show the high potential of our process route [4] for the production of novel polymer blend particle systems for SLS.

References:

[1] I. Gibson, D. Rosen, B. Stucker, Additive manufacturing technologies, 2. ed., Springer, New York, NY, 2015.

[2] T.T. Wohlers, Wohlers Report 2016, Wohlers Associates, Fort Collins, Col., 2016.

[3] D. Drummer, K. Wudy, F. Kühnlein, M. Drexler, Polymer Blends for Selective Laser Sintering: Material and Process Requirements, Phys. Procedia. 39 (2012) 509–517. doi:10.1016/j.phpro.2012.10.067.

[4] J. Schmidt, M. Sachs, C. Blümel, B. Winzer, F. Toni, K.-E. Wirth, W. Peukert, A novel process route for the production of spherical LBM polymer powders with small size and good flowability, Powder Technol. 261 (2014) 78–86. doi:10.1016/j.powtec.2014.04.003.

[5] J. Schmidt, M. Sachs, S. Fanselow, M. Zhao, S. Romeis, D. Drummer, K.E. Wirth, W. Peukert, Optimized polybutylene terephthalate powders for selective laser beam melting, Chem. Eng. Sci. 156 (2016) 1–10. doi:10.1016/j.ces.2016.09.009.

506658 Creating Tuneable Agglomerates via 3D Printing

Abstract:

It is essential to offer a universally accepted way to systematically evaluate agglomerate breakage propensity and mechanisms. Discrete Element Method (DEM) simulation is a commonly used tool but is limited by the lack of identical, controlled agglomerates to test and validate simple models, let alone replicate the complex structure of real industrial agglomerates.

Our research presents a systematic approach to produce 3D printed agglomerates with tuneable properties. Agglomerates with different mechanical properties and structures were designed and printed by an Objet500 Connex 3D printer. The tensile and shear strength of inter-particle bonds was characterized using Instron tester. Breakage results of agglomerates with simple cubic structures and the design of agglomerates with irregular structures were presented. This 3D printing method will allow more rigorous testing of agglomerate breakage models. Discrete Element Method (DEM) has been used to simulate the agglomerate crushing using the Timoshenko Beam Bond Model (TBBM) with bond properties matching the 3D printed agglomerates. For both two agglomerate structures used in this research, the simulation and experiment showed similar breakage patterns. However, the simulation results significantly underestimated the compressive loads of spherical random structured agglomerates. In order to improve the simulations of compressive granule breakage, the influences of bond geometry definitions and non-linear behaviour of the bond contacts need to be explored.

508699 Effect of Particle Size Distribution and Wettability on Penetration Behavior, Granule Formation and Granule Properties in Single Drop Granulation

Abstract:

In wet granulation, single drop granulation (where one drop forms one granule) can simplify the complexity and reveal the fundamentals of the interaction between the liquid and the powder bed. Combining drop penetration time, granule formation mechanisms and the characterization of granule properties is a novel study to achieve a comprehensive map of the single drop granulation process, which is relevant in particulate applications like pharmaceuticals, detergents, agricultural chemicals and catalysts.

In this work, single drop impact of liquid on a static powder bed was studied to investigate the granule formation mechanism, droplet penetration time, and the characterization of granule properties (morphology, internal structure and content uniformity). Water was used as the liquid binder and two pharmaceutical powders, microcrystalline cellulose (MCC) (mean particle size ~54.8 Î¼m, contact angle with water ~36Â°) and acetaminophen (APAP) (mean particle size ~9.74 Î¼m, contact angle with water ~89°), were mixed to make heterogeneous powder beds with varying compositions. The complete drop impact and penetration was recorded with a high-speed camera. The granule morphology, its internal structure and content uniformity were characterized by the prism method with image analysis, micro-CT and UV-vis spectrometry, respectively. Besides the current MCC and APAP system, MCC and APAP with other particle sizes and relative contact angles will also be studied. It is believed that the mean particle size of the powder bed is predominant in influencing the granule formation mechanism, drop penetration time, and granule properties.

508017 Atomically Deposited Sintering Aids: Assessing the Effects of Alumina Particle ALD on the Sintering and Performance of SOFC Electrolytes and Dental Ceramics

Abstract:

The crystal structure of zirconium oxide (ZrO_2) can be tailored by the incorporation of aliovalent dopants, most commonly, yttrium oxide (Y_2O_3). The concentration of dopant dictates the thermodynamically stable crystal phase where 3 mol%
Y
conducting cubic phase. To achieve suitable mechanical properties and ionic conductivity, Y-TZP and YSZ ceramics must be near theoretical density, requiring high sintering temperatures. Previous researchers have successfully lowered the required Y-TZP and YSZ sintering temperatures through incorporation of aluminum oxide (Al_2O_3) particles. While this technique has proven to reduce the sintering temperature in both systems, the presence of particulate Al_2O_3 inclusions in densified parts reduces the homogeneity of the material which may have deleterious effects on densification and microstructure. In this work, we report on the use of atomic layer deposition (ALD) to conformally coat individual Y-TZP and YSZ particles with the desired amount of Al_2O_3. Constant-rate-of-heating (CRH) experiments were conducted using a horizontal push-rod dilatometer to extract the activation energy of sintering and also gain mechanistic insights into the active diffusion mechanisms as a function of ALD film thickness. Transmission electron microscopy (TEM) and scanning electron microscopy (SEM) were used to characterize the as-deposited thin Al_2O_3 films and the effects of the Al_2O_3 films on grain growth and microstructure during densification. Ionic conductivity and flexural strength tests were conducted to determine the effect of Al_2O_3 thin films on relevant material properties. Al_2O_3 incorporation by ALD is shown to have unique phenomenological effects on the densification of Y-TZP and YSZ as compared with conventional Al_2O_3 incorporation using particle-based approaches.

508644 Effect of Particle Size Distribution and Operating Parameters on Conduction and Convection Heat Transfer Mechanisms in Rotary Drums

Abstract:

Granular materials undergo process steps that include transportation, drying, heating and chemical or physical conversion. In several instances, these materials need to be heated or cooled during processing, and rotary drums (kilns) are the most commonly used process equipment for this purpose. It is not always possible to conduct experiments using different materials and consider all process conditions to study the thermal mechanism due to the inherent complexity of these processes. Therefore, computational modeling and simulations have been extensively used to understand the heat transfer mechanism in rotary drums. To maintain simplicity and to overcome the challenges of simulating polydispersed particles, a common practice is to assume that the polydispersed particles with a narrow size distribution behave like monodispersed particles thermally. However, no effort has been made to validate this assumption. This work focuses on understating the effect of polydispersed particles on the conduction and convection heat transfer mechanisms in indirectly heated rotary drums. It also aims to investigate the effect of rotation speed, fill level, particle size and their interaction effects on the conduction and convection heat transfer mechanisms in rotary drums. For this, the discrete element method (DEM) (using MFIX-DEM, an open-source computational fluid dynamics solver suite) simulation technique is used to analyze the thermal behavior of spherical silica beads, which are the most common catalyst support material.

507974 Selective Hydrogenation of Citral over Supported Pt Catalysts on Various Substrates

Abstract:

Highly dispersed platinum (Pt) nanoparticles (NPs) were deposited on various substrates by atomic layer deposition (ALD) for the reaction of citral hydrogenation. The substrates included multi-walled carbon nanotubes (MWCNTs), silica gel (SiO_2), commercial Î^3-Al_2O_3, and ALD-prepared porous Al_2O_3 particles (ALD-Al_2O_3). Based on TEM and CO-chemisorption analysis, the Pt NPs on all substrates were around 1.3 nm, and Pt/MWCNTs, Pt/ALD-Al_2O_3 and Pt/Î^3-Al_2O_3 showed similar Pt dispersion (75%-85%). But the Pt dispersion on SiO_2 was low (63%). In the reaction of citral hydrogenation, Pt/SiO_2 showed the highest activity due to the strong acidity of SiO_2 and the conversion of citral reached 82% after 12 hr with a selectivity of 58% of unsaturated alcohols (UA), including geraniol and nerol. Among all catalysts, Pt/MWCNTs showed the highest selectivity of UA (65%) due to its unique structure and electronic effect. In the cycling test, Pt/MWCNTs and Pt/ALD-Al_2O_3 catalysts were more stable than Pt/SiO_2 which could be attributed to the different interactions between the catalysts and the solvent.

508345 Numerical Simulation of Particle Classification in a Classifier Based on Coanda Effect

Abstract:

Efficient and precise classification of particles has been an important issue in many industrial applications as well as in academia. A relatively new particle classifier based on 'Coanda effect' was developed and reported to have many favorable features (Liang 2010). They include simultaneous multiple classification, simple design with no moving parts, and easy scale up to higher production rates. However, this classifier has attracted little attention so far, and our understanding of underlying mechanism associated with classification performance needs to be improved.

The purpose of this study is to investigate the detailed fluid flow and particle transport characteristics in a Coanda classifier using CFD (Computational Fluid Dynamics). The main focus is placed on the effects of working conditions (inlet flow velocity, particle diameter) and numerical method for turbulence dispersion of particles on the classification performance. As a numerical method, three different approaches are considered: k-ε model, k-ε model combined with DRW (Discrete Random Walk) model, LES (Large Eddy Simulation). In the first approach, standard k-ε model is adopted to obtain the mean flow and then particle trajectory is calculated based on the mean flow without consideration of turbulence dispersion. However, in the second approach using DRW model, the instantaneous flow velocity is constructed under the assumptions of isotropic turbulence in order to include turbulence effect. Finally, in LES, the instantaneous flow field is dynamically and accurately obtained at each time step. Numerical simulations are carried out for two-dimensional mean flow in case of k-ε model, but for three-dimensional unsteady flow in LES. Particles are tracked using a Lagrangian approach based on one-way coupling. We consider talc particles with density of 2,860 kg/m^3 and diameter of 1 to 20 μm.

The Coanda classifier is mainly composed of three inlet channels, three outlet channels, and Coanda block. Inlet channels are divided into one main inlet and two control inlets. The main air jet containing particles is supplied with the velocity of 35 m/s through the main inlet. Two control air jets with velocities of 32 m/s (lower) and 40 m/s (upper) are also supplied through the control inlets for enhanced control.

The trajectories of moving particles obtained from the mean flow (k-ε model) are analyzed for different diameters. The main jet containing particles deflects toward the curved Coanda block due to Coanda effect as expected. Fine particles (d≤2μm) follow the flow streamlines and enter the lower outlet channel. However, coarse particles (d≥8μm) are detached from the streamlines and go into the upper outlet channel. This observation agrees well with the previous result (Liang, 2010).

Even though the mean flow and corresponding particle trajectories provide useful information, realistic particle passage do not take a form of single curve but they change in time due to flow turbulence. Therefore, particle trajectories are simulated again, turbulence dispersion being considered with DRW model. Due to turbulence dispersion, particles of the same diameter are scattered around the particle pathline obtained from the mean flow. As a result, particles of the same diameter can not be collected into a single outlet. For example, some particles (about 5%) of 1 μm enter the middle outlet even though most of them (about 95%) enter the lower outlet designed for collecting fine particles. This means that flow turbulence deteriorates the accuracy of classification.

Finally, LES simulations are performed to understand dynamic behaviors of fluid flow and particle transport. Instantaneous flow fields such as vorticity contours clearly show unsteady fluid motions inside the classifier. It is also shown that the generation of unsteady vortices leads to the scattering of particle trajectory.

A more detailed analysis including the classification efficiency will be given in the final presentation.

Acknowledgements

This research was supported by the National Research Foundation of Korea (NRF) under grant number 2017M2A8A4018482 and 2015R1D1A1A0105 9675.

References

Liang, S. (2010). *Chem. Eng. Comm.*, **197**, 1016-1032.

503491 Numerical Simulation of the Secondary Air Distribution Layer in the Gas Solid Fluidized Bed Based on Flunt Software

Abstract:

Due to the continuity beneficiation in the practical production, the gangue would accumulate on the distributor forming the gangue layer, which was called Secondary Air Distribution Layer and has vital effect on the gas distribution. In this paper, the ventilation characteristics of the secondary air distribution layer was analyzed. In addition, FLUENT software was used to study the effect of the modeled layer for the fluidity characteristics (e.g., pressure drop, bed height, volatility, etc.) in the gas-solid fluidized bed with the complex layer. The results showed that the complex layer could promote the secondary distribution of the airflow in the bed, which effectively inhibited the production of large bubbles and reduced the bed pressure drop fluctuation. Thus, itâ€™s of significance to provide synergy effect on the fluidization stability in the fluidized bed. Based on the above analysis, the introduced complex structure would be convenient to achieve the coal separation effectively.

507949 CFD-DEM Simulation of the Fluidization of Non-Spherical Particles in Fluidized Bed

Abstract:

In chemical and energy industries, fluidization concerning non-spherical particles (such as rod-like particle, disk-like particles, ellipsoidal particles, etc.) is often encountered, but research in this area is minimal. Yet understanding fluidization of non-spherical particles is crucial in many relevant industrial process. To investigate the fluidization of non-spherical particles, a CFD-DEM coupling algorithm is employed in this research to simulate a bubbling fluidized bed. The non-spherical particles are modeled by super-ellipsoids, multi-sphere models, and other particle models. The simulations results show that, the fluidization of non-spherical particles has significant differences from that of the spherical particles. The local void fraction and particle's projected face area perpendicular to fluid flow direction are directly related to the orientations of the non-spherical particles, which has a significant influence on the particle drag force and fluidization. Due to the significance of particle orientation, we further investigate the factors that influence the distribution of particle orientations. The results show that particle orientation depends on the fluidization time, fluidization velocity, and structure of the fluidized bed.

508320 CFD and DEM Simulation of the Cold Spray Process for Surface Coating with Fine Particles

Abstract:

The coating and functionalization of technical surfaces has been become increasingly important in the last two decades for various applications, since of its manifold possibilities for process optimization, as friction reduction, corrosion resistance or improved wetting behavior. Cold spray is a thermal spraying method for surface coating. It was developed in the late 1980s in Novosibirsk [1] and has gone through continuous optimization on certain applications since then. For example showed our previous study a reduction of the friction properties on steel pins, microstructured with fine Titania particles by an adapted spraying method [2]. In the cold spray process particles are dispersed in a gas stream, which is heated up to several hundred degrees and accelerated to supersonic velocities in a Laval nozzle. The particles impact on a substrate with about 800 m/s and form a strong bonding due to the high kinetic energy which is transferred to deformation and heat in the contact zone. In contrast to other thermal spraying methods, the temperatures occurring during the process are far below the melting temperatures of the used materials, which makes the process suitable also for thermal sensitive powders. The quality of the produced coating layers depends highly on the properties of the powder and substrate as well as on the nozzle design and the process parameters as initial temperature and particle velocities. Due to the high gradients, the high occurring velocities and the small size of the particles, typically in the range of 1 to 50 μm, the contact zone of impacting particles can only be measured indirectly and with destructive methods. Therefore, still not all phenomena which lead to a strong bonding in the contact area and a high quality coating are completely understood. For the study of these mechanisms two different simulation approaches have been used:

a) Simulation of particle and gas flow with Computational Fluid Dynamics (CFD)

For a better understanding of the spray process, the nozzle, the adjacent flow areas and the substrate were computed with a CFD method. The trajectories of single particles were calculated by an Euler-Lagrange approach. For the flow the k-Ï‰ SST model was used. Local temperatures in the substrate were calculated by the law of heat conduction. The impact conditions dependent on the nozzle geometry and the particle properties. With the local temperature distribution in the substrate and in the particle, precise impact conditions could be estimated. A study has been conducted to vary the parameters of the spray process, temperature and pressure as well as the particle properties as shape factor, density and size. Moreover the nozzle geometry has been varied in the simulation to optimize the spray process. The simulation results are compared to images of coated surfaces obtained by Scanning Electron Microscopy (SEM) for validation.

b) Simulation of particle-surface impact behavior with Discrete Element Method (DEM)

The DEM is widely used for investigations of particular processes. The advantage of this method is the detailed calculation of particle-particle or particle-wall contacts to obtain the important parameters as forces and deformations. The contact force in the particle wall contact is also dependent on the surface morphology. Therefore the surface topography of two real technical surfaces, a polished and an unpolished Titanium substrate were measured by Scanning Probe Microscopy (SPM) imaging with a nanoindenter (TI Premier, Hysitron). The surface topography was discretized and implemented in a DEM model. A parameter study has been conducted to determine the distribution of forces and contact angles occurring at the particle impact. Different particle sizes and impact velocities were studied.

The cold pray process dependents on various parameters which affect each other. Since of the high gradients and the small scales occurring during the process experimental measurements are often difficult. With this numerical study different parameters could be considered uncoupled of the other parameters. The results can be used for a better process understanding and an optimization of commercial cold spray systems.

[1] Alkhimov, A. P., Papyrin, A. N., Kosarev, V. F., Nesterovich, N. I., Shuspanov, M. M.: Gas dynamic spraying method for applying a coating; US Patent, 5302414; 1994

[2] Buhl, S., Schmidt K., Sappok D., Merz, R., Godard, C., Sauer, B., Kerscher, E., Kopnarski, M., Antonyuk, S., Ripperger, S.: Surface structuring of case hardened chain pins by cold-sprayed micro particles to modify friction and wear properties; Particuology 21, 2015

508568 Two-Fluid Validation of Constituive Models for the Simulation of Cylindrical Particles

Abstract:

Two-fluid simulations are a popular tool in the modeling of large processes. The constitutive models that close the governing equations are based in granular kinetic theory and typically assume that the particles are spherical. In reality, very few processes utilize spherical particles, hence there is a need for constitutive models that describe more complex shapes. Previous work used discrete element method (DEM) simulations to create a solid stress[1,2] and collisional dissipation rate[3,4] model for cylindrical particles that can replace the current spherical models. These new models have been added to MFiX and this work focuses on their validation. Cylindrical particles of various aspect ratios and solid densities were used in cratering and hopper discharge experiments. Simulations are designed to recreate the experiments in order to compare the results for crater depth and hopper discharge rate. The model effectiveness is assessed with regards to assumptions made concerning particle rotation and preferential alignment.

1. Guo Y, Wassgren C, Ketterhagen W, Hancock B, James B, Curtis J. A numerical study of granular shear flows of rod-like particles using the discrete element method. *J Fluid Mech*. 2012:1-26.

2. Berzi D, Thai-Quang N, Guo Y, Curtis J. Stresses and orientational order in shearing flows of granular liquid crystals. *Phys Rev E*. 2016;93(4):40901.

3. Buettner KE, Guo Y, Curtis JS. Using the Discrete Element Method to develop Collisional Dissipation Rate Models that Incorporate Particle Shape. *AIChE J*.

4. Berzi D, Thai-Quang N, Guo Y, Curtis J. Collisional dissipation rate in shearing flows of granular liquid crystals. *Phys Rev E*. 2017;95(5):50901.

509002 The Comparsion of Coarse Grained CFD-DEM for Simulating the Dense Bubbling Fluidized Bed" Requires Immediate Action

Abstract:

CFD-DEM is an effective tool to study the hydrodynamics in dense bubbling fluidized bed. However, the expansive cost of calculating the particles motion is not easy for its application in industrial application or dense particles system. Thus several coarse grained models (CGM) were put forward to simplifying the calculation of particles and decreasing the particles motion, such as PSM, SPA, MSPA,CGEE, EMMS-DPM where a group of particles are set as a parcel or coarse particle to reduce the computation. In this study the experiment and simulation test conducted to test five coarse grained methods. The result indicated that the models increasing the gravity force, pressure gradient force and drag force linearly with the coarse grained ratio (CGR) can be effective to ensure similar macro behavior between original system and coarse graining system. The simulation data of the solid fraction distribution and bed expansion ratio reflected good agreement as experiment. But the bubbles structure became less clear as increasing the CGR. Furthermore, the power spectrum density (PSD) and Wavelet analysis of the pressure signal were used and showed that SPA and EMMS-DPM gave better agreement result in term of dominant frequency and energy ratio at mesoscale feature. Finally the computational efficiency for different models was compared and the models of EMMS-DPM, PSM and CGEE have higher computation efficiency than SPA and MSPA relatively if the collision model of spring-damping were adopted.

509153 Stability Analysis of Uniform Gas Solids Flow

Abstract:

Uniform distribution of gas solids flow should be ensured to gain sufficient performance of fluidized beds. The highly chaotic entropy dissipation across solids bed will trigger non-uniform fluidization, although distributor plays an important role in obtaining homogenous gas stream. Hence, total solids pressure loading (Î¦$_T$), the pressure drop ratio between solids bed and distributor, is introduced here to present their interaction. The boundary between uniform and non-uniform fluidization can be detected by stability analysis based on Prigogineâ€™s minimum of entropy production principle, and this criterion for *N* parallel paths can be simplified into calculating the second derivative of pressure drop through one path with respect of Î¦$_T$. Furthermore, a phase diagram, illustrating the effects from operational parameter (Î¦$_T$), geometrical characteristics of distributor (C_d) and property of solids (Ga) on instability of uniformity, is attained to draw some useful conclusions of robust uniform fluidization.

509309 Numerical Simulation of Particle Sedimentation Related to Nuclear Safety By CFD-DEM Algorithm

Abstract:

Numerical Simulation of Particle Sedimentation related to Nuclear Safety by CFD-DEM Algorithm

Byoungcheol Hwang[1], Kiyofumi Moriyama[1], and Hyun Sun Park[1*]

[1]Division of Advanced Nuclear Engineering (DANE), POSTECH, Pohang, Korea

**corresponding author: hejsunny@postech.ac.kr*

Abstract

The Discrete Element Method (DEM) is a technique to treat particle-particle interactions, and is used in many industries such as civil engineering, powder processing, pharmacy, and so on. Coupling of this method with Computational Fluid Dynamics (CFD) is needed in problems where the particles interacts with fluid significantly, such as bubble column reactor, mineral processing, and so on.

The simulation of important phenomena in nuclear power plant accidents that involves significant core damage and melting, i.e. so called severe accidents, is also one of the topic where the CFD-DEM is useful. After the Fukushima accident in 2011, the severe accident analysis and preparation of counter measures to mitigate the accident consequences became regulatory requirements in many countries. One of the measures for the cooling and stabilization of the molten core is the flooding of the cavity under the reactor vessel where the dropping core melt is caught in a water pool. It is expected that the molten core breaks up into particles of several millimeter sizes and accumulates on the floor, then makes a particle bed, which is hopefully coolable. In this situation, the assessment for the coolability of the particle bed is very important, and a number of experimental and analytical approached suggest that the internal/external structure of the particle bed affects strongly the coolability. The CFD-DEM is a promising candidate for an analysis method that can clarify such details of the particle bed formation.

We tested the applicability of the multiphase CFD-DEM code, an open-source developed by CFDEM®coupling in 2011, for the simulation of particle sedimentation and particle bed formation under water. Our focus in the present work is on the prediction of the particle falling characteristics and particle bed shape. Although the real condition during a severe accident includes very high temperature particles (>1000K) and boiling water (two-phase flow) in a large scale, we started with a simulation of much simple experiment cases without heat transfer or gas phase, with spherical particles only as the first step for the validation of the method. The experimental data with solid particles of various diameters (4 mm, 6 mm) and densities (3600, 6000, 7800 kg/m^3), particle release nozzle diameters (20, 30, 40 mm) were referred to. The results of numerical simulation were compared with the experimental data in terms of particle falling time, maximum particle dispersion angle, particle average falling velocity, particle bed distribution, and so on. The particle falling time was well predicted by the simulation with the maximum deviation 5%. The maximum particle dispersion angle by the simulation significantly deviated from the experimental data probably due to the influences of initial disturbances. The particle average falling velocity was overestimated. The final particle bed profile by the simulation showed agreed with the experimental observation accurately in the peripheral region, but the height of the center part was underestimated. The fluid dynamic drag model and the particle-particle contact model would need improvement for better accuracy. The present work confirmed the validity of CFD-DEM algorithm on the application for the particle sedimentation in a simple liquid pool. The particle-pool interaction including the gas phase will be the next step.

Keywords: CFD-DEM; particle-fluid interaction; particle sedimentation; nuclear safety; severe accident

509440 Application of Extedned Discrete Element Method to the Melting Process of a Packed of Particles

Abstract:

Application of eXtedned Discrete Element Method to the melting process of a packed of particles

Mehdi Baniasadi, Maryam Baniasadi, Bernhard Peters, Xavier Besseron

Université du Luxembourg, Faculté des Sciences, de la Technologie et de la Communication, Luxembourg

e-mail: mehdi.baniasadi@uni.lu, maryam.baniasadi@uni.lu, bernhard.peters@uni.lu, xavier.besseron@uni.lu, web page: http:/www.xdem.de

ABSTRACT

The phenomena of melting takes place in many industrial processes, such as metal processing, environmental engineering and thermal energy storage systems. The problem of modeling of solid-liquid-gas flow including physical and chemical conversions of particles such as melting is the subject of intense research in industrial processes. These systems are difficult to model with numerical analysis due to the complex interaction between the different fluids and granular phases condering heat, mass and momentum excahnge. Basically models used to predict multiphase flows with particles are Eulerian-Eulerian and Lagrangian-Eulerian. The difference between the two models is based on different treatment of the movement of the particles. The particles are tracked individually in Lagrangian-Eulerian approach while they are considered as a continuum medium in Eulerian-Eulerian. Lagrangian-Eulerian is closer to the real physical processes rather than the Eulerian-Eulerian model for granular flows with particleâ€"particle interaction[1].

In this paper, our main focus is on the modelling of the melting process for a packed bed of particles subject to a fluid flow using a numerical technique based on Lagrangian-Eulerian so-called eXtended Discrete Elemnet Method (XDEM)[2]. In this approach, the particles are resolved as discrete phase coupled via heat, mass and momentum transfer to the surrounding fluid phase. The Discrete Element Method (DEM) coupled by heat transfer is used in order to evaluate temperature distribution, melting rate as well as velocity, oriantion and position for each particle individually. In addition the approach allows to be linked with continues numerical approaches such as Computational Fluid Dynamics (CFD) in order to solve fluid fields. The model has been validated by comparing predicted results with existing experimental data for melting of a single ice particle as well as a packed bed of ice particles in the horizontally flowing water. Fluid-particle and particle-particle intractions has also been considered. In addition, the model has the capability to be extended to the packed bed of particles with different size and properties to produce different liquid phases.

REFERENCES

[1] Dierich, F., Nikrityuk, P. A. & Ananiev, S. 2D modeling of moving particles with phase-change effect. *Chem. Eng. Sci.* **66,** 5459â€"5473 (2011).

[2] B. Peters, Thermal Conversion of Solid Fuels (WIT Press, Southampton, 2003).

514248 Real-Time Interpretation of Chaotic Characteristics and Instability of Dense Gas Solid Fluidization

Abstract:

The establishment of instability monitoring system targets at analyzing the chaotic characteristics and evaluating the instability of the fluidization process for dense gas solid fluidization. Experimental pressure fluctuation and pressure drop measurements were interpreted from dense gas solid fluidization system based on chaotic time-series analysis. The pressure fluctuation signals were filtered in terms of attenuating the noise signals. The filtered pressure signals were analyzed with respect to attractor reconstruction. Based on the reconstructed attractor, the maximum Lyapunov exponent and entropy were investigated and estimated under different bed heights and air velocities. Results show that the maximum Lyapunov exponents udner different bed heights and air velocities are positive, which indicates that dense gas solid fluidization exhibits chaotic characteristics. With the increase of bed height, estimated entropy of dense gas solid fluidization increases with the increase of bed height under identical air velocity, while it varies slightly with the increase of air velocity when the air velocity alters from minimum fluidization velocity to 1.36 times of the minimum fluidization velocity.

508510 Glucan Particles As Potential Carriers of Natural Flavonoids for Treatment of Idiopatic Inflammation Diseases

Abstract:

One of the novel methods of pharmaceutical formulation is the loading of the drug into biocompatible microparticles, preferably extracted from natural sources. This approach offers several potential advantages, *e.g.* increased bioavailability (the ratio of the administered drug, which actually enters the bloodstream) compared to conventional drug delivery methods, enabling the option of oral administration even for parenteral drugs. Multiple principles of the loading process are possible – drug encapsulation, adsorption onto the particle surface or precipitation inside the pores a hollow particle. The presented work focuses on precipitation of a drug from an organic solvent inside glucan particles. Glucan microparticles are ellipsoid porous shells, which are obtained from cell walls of *Saccharomyces cerevisiae* (baker´s yeast). The particle size is 2-5 micrometers. Cell walls of glucan particles are mainly composed of β-1,3-glucans. In order to prevent inflammatory reaction in organism, the yeast organelles are removed by a series of alkaline and acidic extraction steps. Subsequently, present lipopolysacharides are extracted using different organic solvents.
Lymphatic cells are mainly present in inflammation area in organism where they are responsible for progress of inflammation. Glucan particles (GPs) interact with lymphatic cell´s receptors on their surface and after that β-1,3-glucans are phagocytosed by the lymphatic cells. Therefore, GPs represent an effective oral delivery system for transporting of biological active substances. Curcumin is an anti-inflammatory flavonoid, which occurs in plants of family *Zingiberacaea.* In our project, curcumin was delivered to an inflammation in organism using GPs. Loading of curcumin into GPs was enabled via precipitation of curcumin from organic solvents. Curcumin dissolved in organic solvents was added to the GPs and through pores penetrated into cavity of glucan shell. Next, the solvent was removed and the curcumin was precipitated in glucan microparticles. GPs with the flavonoid inside were observed using confocal microscopy and SEM (scanning electron microscopy). The achieved content of curcumin was successfully analyzed by other methods, which can not be mentioned directly due to article in preparation. Furthermore, the curcumin-loaded GPs were tested *in vitro* on macrophage like cells and *in vivo* on laboratory mice in order to evaluate their effectivity. The obtained results show that these composites are promising materials for biomedical applications such as efficient treatment of inflammatory diseases (e.g. idiopatic inflammatory bowel disease, rheumatoic arthritis).

508601 Use of Advanced Imaging Techniques in Tablet Disintegration Study

Abstract:

Certified testing of tablet disintegration and dissolution provides us useful information. However, there are many more that are hidden due to techniques' restrictions. Such information is e.g. mechanism of disintegration, change of surface morphology or uniformity of the substances' distribution in the tablet. Therefore, there are many techniques used to obtain the remaining information such as Magnetic Resonance Imaging (MRI), Fourier Transform Infra-Red (FTIR) or Raman spectroscopy and microscopy, Terahertz spectroscopy, Scanning Electron Microscopy (SEM) and other.

Presented work uses tablet made by direct compression of binary mixture of ibuprofen and lactose (based on previous results there was 0.2 - 0.5 w/w of ibuprofen content used) pressed at different compaction pressures (low, medium and high). The tablets were subjected to several solutions and were studied by techniques that have insight into the tablet (MRI) and ones that can see the surface (FTIR and SEM).

MRI experiments provided us with several interesting and several expectable results. As the content of ibuprofen was increasing so was increasing the dissolution time and the wetted layer was decreasing. Similar results were observed when the compaction pressure was increased. When the tablets were submerged to dissolution limiting solutions (saturated lactose solution and solution saturated with both lactose and ibuprofen) the disintegration was severely limited and some tablets began to increase their volume. However, in case of 0.2 w/w ibuprofen tablets it didn't matter whether there was dissolution discriminating solution or not, it disintegrated anyway.

FTIR microscopy showed, that not only the amount, but also the shape and size of the particles from which the tablets were compacted can be determined. These findings were further supplemented by the dry state information obtained by Scanning Electron Microscopy after different dissolution times. Further the FTIR-ATR imaging technique helped to understand the basic mechanisms of the tablet disintegration and to study the dissolution and water penetration into the model ibuprofen-lactose tablets.

508632 Correlation Study between Liquid Penetration and Mechanical Properties of Pharmaceutical Tablets

Abstract:

Main phenomenon influencing treatment of patient in pharmacy is amount of active pharmaceutical ingredient dissolved in medium such as water or blood. Even though this isa key step to achieve desirable medical effect it is not the first one occurring.The first step is governed by surface properties such as wettability and roughness and the first contact point is followed by liquid penetration which is highly dependent not onlyon type of used liquid, ionstrength, liquid density but also on mechanical and structural properties of tablets such as porosity, structure strength or particle bonds.Through many analytical methods we can describe each of thesesteps individually. Aim of this study is to find proper correlation amongseveralmethods such as sessile drop method, texture analysis (TA) and magnetic resonance imaging (MRI) of tablet during dissolution.

For experiments binary mixture of tabletwas used. Tablets consist of lactose (hydrophilic) and ibuprofen (hydrophobic)with variable content (from 0.2 to 0.5 w/w ratio) and were also variable in compaction pressure used on them (from low â€“ 5 kN through 10kN as middle to 18 kN as high).Three different solutions were used according to their saturation with primary substances to prevent the tablet from dissolving. Surface properties were evaluated on contact angle and soaking rate of liquid, both measured by sessile drop method. Texture analysis was performed by recording the material resistance during dry and wet state of tablet. Then the data was processed and trends were obtained.Tablets during the MRI experiments were submerged into dissolution medium and the proceeding wetting front was measured. Later the images were processed via image analysis and trends were obtained.

Results obtained from surface experiments show that higher the ibuprofen content is the more hydrophobic surface appears to be and therefore the contact angle is increasing. Contact angle is independent on manufacturing pressure but soaking rate on the other hand depends strongly.MRI images show that this technique is not suitable for rapidly disintegration tablets, where the calculated error is within experiment conditions. On the other hand, tablet with prolonged dissolution are well measured. Results from MRI experiments also state that local saturation with substances supress dissolution of tablet. Texture analysis of tablets show that they undergo immediately structural weakening after being poured with solution independently on type of solution. Rate of this structural change mainly depends on solution's acidity. Thenthe tablet decreases its resistance in time which can be interpreted as relaxation rate. With increment of ibuprofen or saturation with primary substances this rate decrease, which means that dissolution is suppressed. Correlation between MRI and relaxation rate was done. Results shows that the correlation is mainly dependent on texture analysis measurement. On the other hand, correlation between MRI and soaking rate appears to be more tentative.

Experiment characterising surface, structural and liquid penetration properties were investigated by three different techniques. Obtained results were evaluated and put to mutual correlation. From liquid penetration point of interest there is correlation between MRI and contact angle method. Also, a correlation between MRI and texture analyser was found.

508642 In Situ Drug Amorphisation By Microwave Irradiation Stabilized By Mesoporous Silica

Abstract:

One of the common issues in the pharmaceutical industry is the poor solubility of the newly developed active pharmaceutical ingredients (APIs). The solubility might be significantly enhanced by changing the form of the API from crystalline to amorphous form. The amorphous state is characterized by lower thermodynamic stability and thus better dissolution properties leading to better bioavailability. In addition, the amorphous API in a dosage form (e.g. tablets) must also remain stable and not recrystallize for at least the shelf life of the drug product (usually 2 - 3 years). Various methods are being utilized to prepare and stabilize amorphous formulations in the industry, however they usually involve difficult formulation steps or suffer from poor efficiency, low stability or can be patent obstructed.

In this work a stable amorphous formulation is prepared simply using the standard wet and dry granulation process or direct tableting. Crystalline form of API is used during the initial process. The amorphous form is then prepared *in situ* and â€œon demandâ€ either during manufacturing, or by the end user by applying microwave radiation (MW).

To stabilize the API in an amorphous form after the MW, multi-structured mesoporous silica particles were added to the formulation. The particles were prepared using an emulsion synthesis of octylamine in tetraethyl orthosilicate. Using ethanol as an inhibitor instead of the commonly used nitric acid as a catalyst reduces the silica primary nano-particles nucleation rate. The formed nuclei grow larger and then self-assemble around the octylamine droplets and form hollow micro-particles. The prepared microparticles were characterized using the Scanning electron microscopy (SEM) and the Transition electron microscopy (TEM). To characterize the porous structure, nitrogen sorption (BET) analysis was conducted. The results clearly show the particle assemblies contain macropores (cavities), mesopores and micropores and possess very high specific surface area and also very large pore volume. Together with the thermal and chemical stability of silica these properties make the particles well suitable for preserving the melted API in an amorphous form after the MW.

As a proof of concept, initially powder mixtures and later simple tablets composed of Ibuprofen (model API), silica micro-particles and a filler excipient were experimentally investigated. The *in situ* amorphization of ibuprofen was conducted using a commercial microwave oven. The heating properties of the oven, i.e. the spatial distribution of microwave heating was characterized. Various microwave power input and varying composition of the mixtures (excipient/Ibuprofen ratio) were tested with respect to the melting point of the Ibuprofen. The change from the crystalline to amorphous structure due to MW heating was analyzed by x-ray diffraction (XRD), differential scanning calorimetry (DSC) and FTIR-ATR techniques. The dissolution profiles of ibuprofen from both MW untreated and treated tablets were measured using the High Performance Liquid Chromatography (HPLC).

Based on these initial results, fully formulated tablets were prepared with various API dosage strength using a tablet press. Same compression force and same dwell time was used for all of the tablets. The final tablet composition includes excipient capable of absorption of the MW, which ensures a uniform heating in the whole volume of the tablet. Next it contains the silica microparticles, which by capillary forces absorb the API once melted and stabilize it after cooling down.

Lastly, the tablets contain standard filler, lubricant and disintegrant excipients. After the MW application the tablets change their structure due to the melted API intake into the silica micro-particles, forming visible pores in the tablet structure. This has a notable impact on the dissolution and hardness of the resulting tablets. For the final tablet characterization SEM, Texture Analysis, FTIR and Magnetic Resonance Imaging were used.

508646 Manufacture of Personalised Medicines By API Printing on Porous Tablets

Abstract:

Personalised medicine offers an opportunity to produce tailor-made medicaments for patients. In order to realize this vision, it is necessary to find an alternative approach to the conventional tablet design and production. One that will replace the existing mass production of drugs offering only a few dose options of the active pharmaceutical ingredient (API), while at the same time it must meet the strict reproducibility and a defined relationship between composition and release rate of active substances requirements.

One of such approaches is to print the API on demand on a solid substrate directly when it becomes needed. Appropriate dosing of API and the possibility to even combine several APIs into a single medicament improves the variability of the dosages available for prescription and reduces the overall number of pills the patient has to take. The goal of this work is to find a simple, automated and cost-effective method of preparing personalised medicaments, which could take place directly in the pharmacy shops or at home.

The method in this work consists of a precise dosing of a dissolved API into an already prepared substrate â€“ porous placebo tablets with high sorption capacity. The tablets are prepared using direct compression in a tablet press from a simple physical mixture of a binder, filler, disintegrant and lubricant. As a filler, mesoporous silica particles were used. The high pore volume provides the free space for API sorption into the tablets and the small size of the mesopores ensure the API is deposited in an amorphous form.

The prepared particles used in the formulation were characterized using Transmission electron microscopy (TEM), Dynamic light scattering (DLS) and nitrogen sorption analysis (BET). The prepared tablets hardness and uniformity was characterized both prior and after loading. The tablets after loading were characterized using the X-ray powder diffraction and the Fourier transform infrared spectroscopy to determine the API form and deposition uniformity, followed by dissolution tests to confirm the applied dosage strength and to measure the release profile. So far, several BCS Class II APIs were successfully infused into the substrates using precise on-demand dosing. Impregnation of various doses of API were accurate. Later characterization revealed no crystalline API were present and the dissolution profiles were very satisfactory.

509441 Understanding Phase Transition of Acetaminophen in the Bulk and Surface of Acetaminophen

Abstract:

Pharmaceutical solids processing often lead to the generation of new phases in the bulk and on the surface of the processed drugs. Phase transformation of the API can potentially change the physiochemical properties and influence the drug performances including solubility, dissolution and stability. To detect these changes, analytical tools such as XRD are usually employed. However, these techniques remain suitable for bulk analysis. Whereas, phase transition on the surface remains hard to detect due to the lack of literature, researches in this area and the absence of surface analytical techniques. The aim of this work is to investigate different phases (pure crystalline, partially crystalline partially amorphous and amorphous) of Acetaminophen drug by combining bulk and surface analytical techniques. For bulk analysis, FTIR and XRD have been employed. The former shows no significant differences between Acetaminophen various phases. Whereas the latter confirmed the presence of different phases indicated by a remarkable variation in the absorbance intensity the peak width. Where, high and sharp peaks obtained for the pure crystalline Acetaminophen, fewer peaks observed for semi crystalline Acetaminophen and a single broad peak observed for the amorphous Acetaminophen. For the surface analysis, Nitrogen K near-edge X-ray absorption fine structure (NEXAFS) spectra of a series of NEXAFS indicates that the precise determination of IPs and Ïƒ* shape resonance energies is sensitive enough to distinguish between the different phases of Acetaminophen. DFT calculations provides information on the molecular orbitals involved during electronic excitation occurred as a result of NEXAFS experiment.

509599 Treatment of Cystinosis through Vitamin E Modified Silicone Hydrogel

Abstract:

Cystinosis is an orphan disease caused by a genetic mutation that leads to deposition of cystine crystals in many organs including cornea. Ophthalmic manifestation of the disease can be treated with hourly instillation of cysteamine eye drops. The hourly eye drop instillation is tedious to the patients leading to poor compliance and additionally, significant degradation of the drug occurs within one week of opening the bottle, which further complicates this delivery approach. Our work focuses on designing a contact lens to treat the disease with improved efficacy compared to eye drops. Our goal is to design a lens that is safe and that can deliver a daily therapeutic dose of cysteamine to the cornea while retaining drug stability. We show that cysteamine diffuses out rapidly from all lenses due to its small size. The rapid release of cysteamine and other ophthalmic drugs can be prolonged significantly by integration of vitamin E aggregates inside of the polymer matrix of silicone hydrogels. The increase in release duration increases as square of the volume fraction of the loaded vitamin E fraction which leads to high efficacy and suggests that the mechanism of attenuation in drug release is formation of nanosized barriers. The increase in duration depends on the type of the contact lens. The barriers are not effective in single-phase hydrogel lenses while the effect in the bi-phasic silicone hydrogels depends on the type of the contact lens. Vitamin E incorporation increases the release duration of both ACUVUE®OASYS® and ACUVUE® TruEye™ but the effect is more pronounced in TruEye™ likely due to the low solubility of vitamin E in the lens matrix and higher aspect ratio of the barriers. This result along with the high aspect ratio seen through SEM imaging in silicone hydrogels suggests that barriers could be forming at the interface of the silicone and hydrogel phases. The presence of vitamin E has an additional beneficial effect of reduction in the oxidation rates, likely due to a transport barrier between the oxygen diffusing through the silicone channels and drug located in the hydrogel phase. Based on this study, both Acuvue®OASYS® and ACUVUE® TruEye™ can be loaded with vitamin E to design a cysteamine eluting contact lenses for effective therapy of cystinosis. The lenses must be worn for about 4-5 hr. each day, which is less than the typical duration of daily-wear. Vitamin E loaded contact lenses could be very useful for treating many other ophthalmic diseases as well.

517683 Systems Integration for Dry Granulation Based Continuous Pharmaceutical Tableting

Abstract:

Over the past decade, the pharmaceutical industry has seen significant progress in the mechanistic understanding of relevant unit operations, availability of multiple sensor technologies for inline monitoring of critical quality attributes and advances in data analytics to support manufacturing decision-making. These have enabled a systematic realization of continuous manufacturing for pharmaceutical tablets using the direct compaction route. However, the limitations and difficulty in handling of some particulate materials provide the impetus for investigating granulation routes to the continuous manufacturing of tablets.

In this work, the integration of dry granulation into a continuous tableting line is investigated. Specifically, the Process Analytical Technology (PAT) tools required for inline and at-line monitoring of the tableting line, and the semi-mechanistic models for the roller compactor and the tablet press are discussed. While the Kawakita model is used for the tablet press, the roller compaction process model is consisting of an integration of material balance and actuator dynamics from Hsu et. al (2010) with the modified Johansonâ€™s model presented by Liu & Wassgren (2016). The PAT tools, mechanistic equations, and automation system are integrated for measurement validation and data accuracy using data reconciliation in a structured ISA-95 network using DeltaV. The experimental data is presented from the operation of the continuous dry granulation line-based pilot plant at Purdue University.

509248 Modeling Granular Material Mixing and Segregation Using a Multi-Scale Model

Abstract:

A multi-scale model is presented for predicting the magnitude and rate of powder mixing and segregation in blenders. The model combines diffusion and percolation correlations from small-scale discrete element method (DEM) simulations with advective flow field information from large-scale finite element method simulations. The multi-scale model predictions closely match results from both large-scale DEM-only simulations and experiments, but they are much more computationally efficient than DEM-only simulations. Therefore, in contrast to DEM, the multi-scale model is well suited for predicting blending in industrial-scale systems.

Modelling Granular Media using Dynamical Density Functional Theory

T. Hurst, B. Goddard & R. Ocone

Motivation

Granular media is ubiquitous in industrial and physical problems. One of the characteristic qualities of granular media at the microscopic level is inelastic hard body collisions, which drive macroscopic behaviour. Modelling granular media is a complex and interesting problem. Solid-like phenomena make it difficult to approximate a granular system with a continuum model, while chaotic behaviour and inelastic collapse in hard particle systems introduce modelling challenges on the microscopic scale. We investigate uses of dynamical density functional theory (DDFT), augmented with data gained from microscopic models, for granular media simulation.

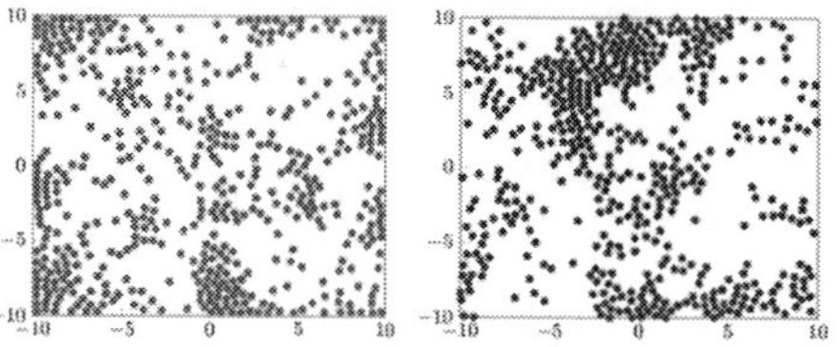

Figure 1: Inelastic systems with reflective (left) and periodic (right) boundaries

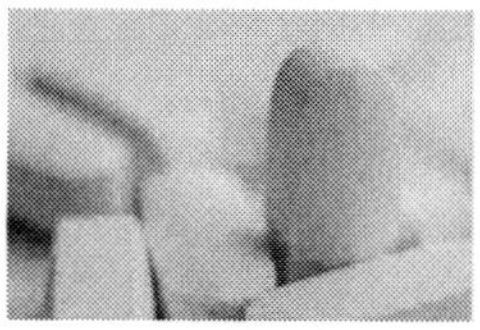

Dynamical Density Functional Theory

To derive a DDFT, we start with a set of Langevin equations for N hard particles of diameter σ in d dimensions with positions $\boldsymbol{r}^N$ and momenta $\boldsymbol{p}^N$:

$$\frac{d\boldsymbol{r}^N}{dt} = \frac{\boldsymbol{p}^N}{m}, \quad \frac{d\boldsymbol{p}^N}{dt} = -\nabla_{\boldsymbol{r}^N} V(\boldsymbol{r}^N, t) - \gamma \boldsymbol{p}^N + (m k_b T \gamma)^{1/2} \boldsymbol{G}(t), \tag{1}$$

on the hard sphere domain, $\mathcal{D}_N = \{(\boldsymbol{r}^N, \boldsymbol{p}^N) \in \mathbb{R}^{2dN} : \forall i, j, \|\boldsymbol{r}_i - \boldsymbol{r}_j\| \geq \sigma\}$, where $V(\boldsymbol{r}^N, t)$ includes external and interparticle potentials, and $\boldsymbol{G}(t)$ is an optional stochastic force term. The Kramer's equation is then a partial differential equation for its distribution function $f^{(N)}(\boldsymbol{r}^N, \boldsymbol{p}^N, t)$:

$$\frac{\partial f^{(N)}}{\partial t} + \frac{1}{m}\boldsymbol{p}^N \cdot \nabla_{\boldsymbol{r}^N} f^{(N)} + \nabla_{\boldsymbol{r}^N} V \cdot \nabla_{\boldsymbol{p}^N} f^{(N)} = \gamma \nabla_{\boldsymbol{p}^N} \cdot \boldsymbol{p}^N f^{(N)} + \gamma k_b T \nabla^2_{\boldsymbol{p}^N} f^{(N)}. \tag{2}$$

We integrate out all but one position and momenta in the Kramers equation, to find:

$$\frac{\partial}{\partial t} f^{(1)} = \mathcal{L}_1 f^{(1)} + \mathcal{L}^{2,\dots,N}_{\text{part}}(f^{(k)}) + \mathcal{L}^{(1)}_{\text{diss}} f^{(1)} + \mathcal{L}_{\text{coll}}(f^{(2)}), \tag{3}$$

where $\mathcal{L}_1$ introduces effects from external potentials, $\mathcal{L}^{(1)}_{\text{diss}}$ effects due to dissipation and friction, $\mathcal{L}^{2,\dots,N}_{\text{part}}$ effects from interparticle potentials, and

$$\mathcal{L}_{\text{coll}}(f^{(2)}) = -\sigma^{d-1} \int_{\mathbb{R}^d} d\boldsymbol{p}_1 \int_{\mathbb{R}^d} d\boldsymbol{p}_2 \int_{S_1^{d-1}} (\boldsymbol{p}_1 - \boldsymbol{p}_2) \cdot \omega f^{(2)}(\boldsymbol{r}_1, \boldsymbol{r}_1 + \sigma\omega, \boldsymbol{p}_1, \boldsymbol{p}_2, t) d\omega, \tag{4}$$

is a general collision term that arises due to the hard-sphere domain, when assuming only binary collisions. By taking the first two $\boldsymbol{p}$-moments and upon additional assumptions we arrive at a closed set of equations to model the dynamics of the system, a DDFT [1, 2]:

$$\frac{\partial}{\partial t}\rho = -\nabla \cdot (\rho \boldsymbol{v}), \tag{5}$$

$$\left(\frac{\partial \boldsymbol{v}}{\partial t} + \boldsymbol{v} \cdot \nabla \boldsymbol{v}\right) + \gamma \boldsymbol{v} = -\frac{1}{m}\nabla \frac{\delta \mathcal{F}[\rho]}{\delta \rho} - \sigma^{d-1} g_\sigma \int_{S_1^{d-1}} \rho(\boldsymbol{r} + \sigma\boldsymbol{\omega}/2)\rho(\boldsymbol{r} - \sigma\boldsymbol{\omega}/2)$$
$$\times \boldsymbol{\omega} \cdot \left\{\frac{k_B T}{m}\mathbf{1} + \boldsymbol{v}(\boldsymbol{r} + \sigma\boldsymbol{\omega}/2) \otimes \boldsymbol{v}(\boldsymbol{r} + \sigma\boldsymbol{\omega}) - \boldsymbol{v}(\boldsymbol{r} + \sigma\boldsymbol{\omega}/2) \otimes \boldsymbol{v}(\boldsymbol{r} - \sigma\boldsymbol{\omega}/2)\right\} \mathrm{d}\boldsymbol{\omega}, \tag{6}$$

where $\mathcal{F}$ is the well-studied Helmholtz free energy operator, which can include particle-particle interactions and external potentials, ρ is the density, $\boldsymbol{v}$ is the local velocity and g_σ is the value of the correlation function at σ, which must be determined. We choose to model g using statistical data from microscopic simulation.

DDFT Simulations

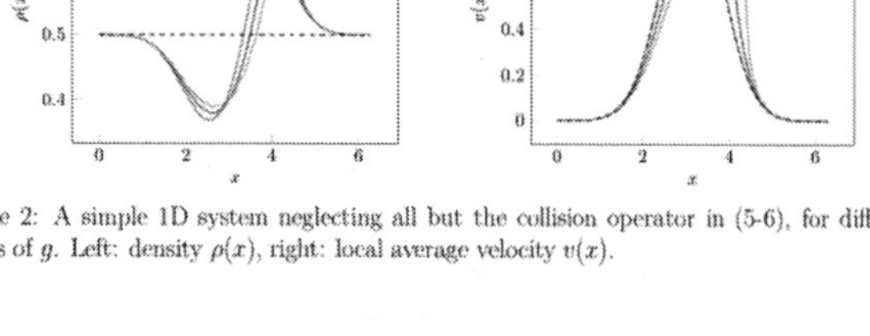

Figure 2: A simple 1D system neglecting all but the collision operator in (5-6), for different values of g. Left: density $\rho(x)$, right: local average velocity $v(x)$.

Data-Driven Modelling of g

A system of N hard spheres of diameter σ in d dimensions with positions $\boldsymbol{x}^N$ and velocities $\boldsymbol{v}^N$ obey

$$\frac{d\boldsymbol{r}^N}{dt} = \boldsymbol{p}^N, \quad \frac{d\boldsymbol{p}^N}{dt} = V(\boldsymbol{r}^N, t), \tag{7}$$

with a pairwise collision condition: if particles i and j are in contact, i.e. $\|\boldsymbol{r}_i - \boldsymbol{r}_j\| = \sigma$ and $(\boldsymbol{r}_i - \boldsymbol{r}_j) \cdot (\boldsymbol{p}_i - \boldsymbol{p}_j) < 0$, we instantaneously change the velocities via the rule:

$$\boldsymbol{p}_i^{out} = \boldsymbol{p}_i^{in} - \frac{1+\alpha}{2}\boldsymbol{\nu}_{ij} \cdot (\boldsymbol{p}_i^{in} - \boldsymbol{p}_j^{in})\boldsymbol{\nu}_{ij}, \quad \boldsymbol{p}_j^{out} = \boldsymbol{p}_j^{in} + \frac{1+\alpha}{2}\boldsymbol{\eta}_{ij} \cdot (\boldsymbol{p}_i^{in} - \boldsymbol{p}_j^{in})\boldsymbol{\eta}_{ij}, \tag{8}$$

where $\boldsymbol{\eta}_{ij} = \frac{\boldsymbol{r}_i - \boldsymbol{r}_j}{\|\boldsymbol{r}_i - \boldsymbol{r}_j\|}$ and $\alpha \in [0, 1]$ is the coefficient of restitution. For simple V, the Newton equations can be modelled using Event Driven Particle Dynamics (EDPD). The cell method and bounded increasing priority queues [3] reduce the computational effort to $\mathcal{O}(1)$ per collision. Using statistical methods for synthetic data (e.g. emulation [4]) we gain an understanding of the correlation function g near equilibrium, which we can include in (6-7).

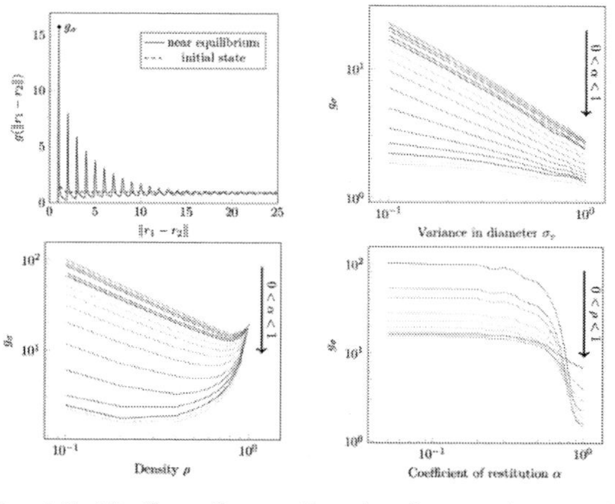

Figure 3: Top left: $g(\|r_1 - r_2\|)$ near equilibrium for a 1D system with $\rho = 0.5$ and $\alpha = 0.5$, using 10000 samples of EDPD simulations. The other three graphs show how g_σ changes for different parameters.

Future Work

We wish to include additional moments in the DDFT derivation, simulate DDFT in higher dimensions including interparticle potentials, and use EDPD in higher dimensions to calculate g.

[1] Archer, A., *J.Chem.Phys.*, **130**, 014509, (2009).
[2] Goddard, B. et. al., *Phys.Rev.Lett*, **109**, 1-5, (2009).
[3] Paul, G., *J.Comp.Phys.*, **221**, 615-625, (2011).
[4] O'Hagan, A., *Reliab.Eng.Syst.Saf.*, **91**, 1290-1300, (2006).
The support from EPSRC (grants EP/N034066/1, EP/L025159/1 and EP/L016508/01) is kindly acknowledged.

509555 A Study on Partially Wetted Particle Collisions with a Wet Wall

Abstract:

For the production of poly-ethylene, a gas-phase polymerization reaction is executed in a fluidized bed reactor. The polymerization reaction produces an excess of heat and the reactor is unable to efficiently remove the excess heat. The introduction of an Inert Cooling Agent (ICA) solves the temperature problem but also has a downside. Introducing the ICA in the reactor influences the particle collisions, leading to substantially different flow behavior. Besides the reduction of the restitution coefficient, the introduction of the liquid can result in adhesion instead of collision. This adhesion results in the formation agglomerates in the reactor and reduces the efficiency of the process. Due to the complex physics of this problem, Computational Fluid Dynamics (CFD) is a suitable method to gain insight in this phenomenon.

In order to study this complex system of partially wetted particle collisions a Direct Numerical Simulation (DNS) is used for incompressible flow. To model the coupling between the solid phases and the fluid phase, a second order accurate Immersed Boundary Method (IBM) is used (Deen et al.,2012). The interface between the fluids is tracked by a Volume Of Fluid (VOF) method (Baltussen et al., 2017). The contact-line dynamics are solved by imposing a contact angle as a boundary condition (Patel et al., 2017).

The simulated results will be compared with dry particle collisions on a wet wall (Tang et al., 2017) and with wet particle collisions on a wet wall.

References:

1. Deen, N.G., Kriebitzsch, S.H., van der Hoef, M.A. and Kuipers, J.A.M., Direct numerical simulation of flow and heat transfer in dense fluidâ€"particle systems. Chemical Engineering Science, 81 (2012): 329-344.

2. Baltussen, M. W., Segers, Q. I. E., Kuipers, J. A. M., and Deen, N. G., Cutting bubbles with a single wire. Chemical Engineering Science 157 (2017): 138-146.

3. Patel, H.V., Das, S., Kuipers, J.A.M., Padding, J.T. and Peters, E.A.J.F., A coupled Volume of Fluid and Immersed Boundary Method for simulating 3D multiphase flows with contact line dynamics in complex geometries. Chemical Engineering Science, 166 (2017): 28-41.

4. Tang, Y., Buck, B., Heinrich, S., Deen, N.G., Kuipers, J.A.M., Interface-resolved simulations of normal collisions of spheres on a wet surface. American Institute of Chemical Engineers, (2017).

509518 Understanding the Role of Drying History on the Crystallisation of Dehydrating Lactose Particles

Abstract:

Spray drying is one of the most commonly used technology to produce powders from atomised droplets that undergo evaporation. Operating temperature is well established to be a crucial parameter in affecting the physio-chemistry of particles, notably on particle crystallinity which provides better stability than amorphous particles. Lactose is a common pharmaceutical excipient. Previous reports suggested the use of high inlet temperature, up to 210°C to improve lactose crystallinity.[1] However, there are reports claiming that lower inlet temperature at 170°C is more inclined to lactose crystallisation.[2] In both cases, detailed study on the drying history and the state of particle throughout the drying time frame have not been explored. Spray drying temperature can only be controlled at the inlet/outlet of spray dryer leading to a constraint in drying temperature variation. This research addresses the current limitation by introducing a modified spray dryer known as 'narrow tube dryer' which entails 12 narrow copper tubes with internal diameter of 1.7 cm. Heat is supplied through the walls of the narrow tube via convection. Each 50 cm tube is isolated, hence enabling the amount of heat supplied to be controlled independently leading to a unique drying profile. With the aid of 1 Dimensional modelling, crucial parameters such as drying temperature profile, particle temperature, moisture content and glass transition temperature can be predicted throughout the drying time frame. This research focus on evaluating the impact of different drying history on lactose crystallinity. Results suggest the existence of an 'active region', a state where molecular movement occurs, as one of the determining factors on lactose crystallinity. The active region can be delineate using the predicted T_p-T_g curves of each drying condition. Within the active region, crystallisation is believed to happen as re-arrangement of atoms and molecules ensues. Particle duration in the active region is crucial, with a longer period providing more time for crystallisation. Lower drying temperature resulted in a shorter particle duration in the active region, hence yielding particle with a lower degree of crystallinity. Though high drying temperature produced a shorter active region, it was found that a sudden drop in drying temperature could instigate crystallisation similar to the crystallisation phenomena occurred in mannitol.[3] This newly developed narrow tube dryer is capable in unveiling the drying history of particle and provide a more comprehensive understanding on particle crystallinity as the information of the state of particle throughout the drying chamber was used to analyse the crystallisation phenomena in spray dried lactose.

Keyword: Crystallisation, Pharmaceutical Engineering & Drug Delivery, Particle Characterisation

1. Chiou, D., T.A.G. Langrish, and R. Braham, *The effect of temperature on the crystallinity of lactose powders produced by spray drying*. Journal of Food Engineering, 2008. **86**(2): p. 288-293.

2. Das, D. and T.A.G. Langrish, *Combined Crystallization and Drying in a Pilot-Scale Spray Dryer.* Drying Technology, 2012. **30**(9): p. 998-1007.

3. Littringer, E.M., R. Paus, A. Mescher, H. Schroettner, P. Walzel, and N.A. Urbanetz, *The morphology of spray dried mannitol particles — The vital importance of droplet size.* Powder Technology, 2013. **239**: p. 162-174.

508979 Scale up Methodology and Practical Application of Wet Bead Milling for the Production of Submicron Particles

Abstract:

Particle size reduction via wet bead milling of a temperature sensitive pharmaceutical material to a reproducible submicron endpoint is achievable across the Netzsch Delta Vita mill series. The reduced particle size allows for an improved dissolution rate when compared to unmilled material. Wet bead milling offers the advantage of increased heat dissipation and removal when compared to a dry milling process. A total energy model approach was used to properly scale the process while geometrically scaling equipment and holding key process variables including formulation composition, agitator tip speed, and bead characteristics, among other parameters, constant between scales. An advantage of scaling with constant tip speed is that the power and heat transfer surface area also scale by the diameter of the mill which results in similar cooling across scales thereby avoiding unwanted surprises for temperature sensitive products during scale up. The milling process was smoothly scaled from lab to industrial scale.

509526 Design and Optimization of an Equipement for Active Pharmaceutical Ingredients

Abstract:

This paper presents the design optimization procedure of a wet bead mill for active pharmaceutical ingredients by highly efficient numerical and experimental methods. The setup of an experimental particle image velocimetry (PIV) system and the data acquisition process are presented. The PIV measurements were used to calibrate and validate a computational fluid dynamics (CFD) simulation model, which is also described in detail, allowing thus to obtain useful and reliable insights on the characterization of the three-dimensional flow inside the mill. This enabled, in a first step, the optimization of specific elements such as the stator or rotor stage geometry of the mill. Furthermore, a ceramic prototype of the rotor was produced while respecting manufacturing and contamination constraints; it demonstrated experimentally that the efficiency of milling had been improved. Secondly, the simulations focused on the filter-free separation of the beads from the ground material suspension. The separation process was optimized numerically until obtaining a result of 100% separation. The interaction between the separation simulations and empirical tests was carried out in a particularly productive manner thanks to the fast prototyping of the separator elements using 3D printing, which confirmed the excellent quality of the separation.

508580 Designing and Characterizing Chitosan Particles for Drug and VaccineÂ Delivery Applications

Abstract:

Chitosan is a nontoxic, biodegradable, and biocompatible polysaccharide useful for several biomedical applications. The purpose of this study is: (1) to use electrospraying, a less widely studied technique, for the generation of micro- and nanoparticles; (2) to develop an empirical relationship between the chitosan particle size and shape and the electrospraying process parameters; and (3) to apply this model for improved encapsulation of bovine serum albumin. The generation of chitosan micro- and nanoparticles allows for the use of a noninvasive, controlled method of administering bioactive agents in drug and vaccine delivery applications. By systematically varying experimental parameters, including voltage, flow rate, needle gauge, chitosan molecular weight, concentrations of cosolvents and surfactants, solution viscosity, and surface tension, the condition under which each particle shape forms is determined. Collected particles are quantified using optical microscopy and light scattering techniques. The microscopy images serve as visual representations of the effects each parameter has on the generation of chitosan micro- and nanoparticles. Preliminary results of this study show how the electrospraying parameters are dependent upon one another. Adjusting one parameter changes the appropriate range of operation of other process and system parameters. Using proper choices of process variables, particles with diameters ranging from the nanometer to micrometer range are obtained. This goal of this study is to develop an empirical model for the production of micro- and nanoparticles for drug and vaccine delivery systems, especially for encapsulating protein-based vaccines in polysaccharide carriers.

KEYWORDS electrospraying, polysaccharide, microparticles, drug delivery

1. Introduction

Electrospraying is a technique that utilizes an electric field to reshape the meniscus of a solution at the tip of a tube, causing the droplets of solution released from the controlled continuous dispensing mechanism to be microscopic in size. Preliminary data have been collected for chitosan solutions that were emitted through the electrospray apparatus to form spherical micro- and nanoparticles of chitosan.

The formation of polyelectrolyte complexes (PECs) with chitosan and bovine serum albumin (BSA) has been evaluated, the results of which will be applied to the electrospraying process. PECs are formed between oppositely-charged polymers due to electrostatic interaction of polyions. Chitosan, a polycation in mildly acidic to neutral conditions,[1,2] aggregates with the model protein BSA due to the anionic nature of the protein under appropriately selected solution pH.

This study focuses on identifying critical process parameters and highlighting important trends in electrospraying so that in the future electrospraying can be a reliable method of creating polymer nanoparticles for drug and vaccine delivery applications.[3]

2. Methodology

2.1 Electrospraying

Chitosan particles were prepared by electrospraying chitosan dissolved in acetic acid. Chitosan solutions of known concentrations were placed in a 10 mL syringe connected to the emitter electrode through silicone tubing. The solution was sprayed from the emitter using a syringe pump that provided a constant specified rate throughout the experiment. At a sufficiently high voltage, the liquid meniscus took the form of an axisymmetric cone, called the Taylor cone. In this cone-jet mode, as the electrostatic forces exceed the surface tension of the solution, a straight, stable jet, < 100 mm in diameter, was formed at the apex of the cone. The end of the jet broke into droplets while traveling toward the collector electrode. The size of the emitted droplets was smaller than that obtained in the absence of the electric field. Different collection media were used throughout the study. The particles collected were quantified using optical microscopy.

2.2 Microparticle synthesis by polyelectrolyte complexation

Solutions of chitosan and BSA were created and adjusted to the desired pH. These solutions were combined in an equal ratio and diluted to be evaluated with dynamic light scattering and UV-vis spectroscopy.

3. Results

3.1. Chitosan Solution Properties

These physical properties of the chitosan solutions were compared with those of 10% acetic acid that served as the solvent.

Molecular weight and pH. The pH of solutions with chitosan concentrations between 1.0 and 2.5 % (w/v) at low, medium and high molecular weights were measured. The molecular weight of chitosan did not have a noticeable impact on the pH of the solution, but the pH increased with the concentration of chitosan.

Surface Tension. Surface tension was an important factor to be analyzed as the chitosan solution must have surface tension low enough to obtain small droplets, but high enough to penetrate the surface barrier of the liquid collection media (yielding a dispersion of microparticles in the collection medium, without forming a film at the surface). The surface tension of each solution was measured using a bubble pressure tensiometer. Overall, the higher the concentration and molecular weight of chitosan in the solution, the higher the surface tension of the solution. To lower the surface tension, ethanol, or other cosolvents or surfactants, were added to the solution.

Ionic Conductivity. The lower the conductivity of a solution being electrosprayed, the more stable the Taylor cone and stable jet is likely to be. As the chitosan concentration was increased, the conductivity of the solution also increased. Molecular weight of chitosan did not have a significant effect on conductivity. Knowledge of the relationship between surface tension and conductivity will not only allow electrospray stabilization, but also make it possible to electrospray smaller and more spherical chitosan microparticles.

Viscosity. The viscosity of the polymer solution is expected to affect both the shape and size of the microparticles The kinematic viscosity of each solution was measured using an Ubbelohde viscometer. Viscosity was found to increase as the concentration and molecular weight of chitosan in the solution increased. It was also found that more viscous solutions, such as 2.5 % of the high molecular weight chitosan, formed a more stable jet at higher flow rates, while less viscous solutions, including 1.0 % of the low molecular weight chitosan and 1.0 % of the high molecular weight chitosan, led to a more stable jet at lower flow rates.

3.2. Effect of Emitter Diameter

Needles of three different sizes (23, 27, and 30 gauge) were used as emitters. After analysis of their influence on electrospraying, the 27-gauge needle was found to be the optimal size for this study and will be used for all future experiments.

3.3 Chitosan-BSA Polyelectrolyte Complexes

Particle size. Prior to electrospraying of chitosan with BSA, particle formation within solutions was characterized. A pH value in the range of 5â€“6, higher than the isoelectric point of BSA (~4.7), was found to be suitable for the desired particle size and particle size distribution. The lower molecular weight chitosan resulted in smaller particles than the higher molecular weight chitosan.

Incorporation of BSA in Microparticles. Centrifugation of the dispersion, and UV-vis spectroscopy of the supernatant solution, was used to determine the fraction of BSA complexed with chitosan in the microparticles. BSA absorbance at 280 nm wavelength was used. More BSA was incorporated in the particles as its concentration in the solution increased. The degree of incorporation was also higher at pH 5 than at pH 6. Up to approximately 47 % of the BSA was found to be complexed with chitosan under these conditions.

The effects of cosolvent/surfactant concentration in the chitosan solution, the flow rate, and the applied voltage on cone-jet formation and the particle size and size distribution, are also under investigation. Taylor cone was not formed below 4 kV, while the jet became unstable above 6 kV.

4. Conclusion

Preliminary results indicate that a limited pH range is ideal for chitosan-BSA complexation. Electrospraying experiments have shown that it is not possible to optimize solution and process variables such as conductivity, viscosity, surface tension, flow rate and voltage independently. Therefore, a combination of properties should be considered for tailoring particle size and shape. Utilizing these optimal parameters will lead to a steady jet and Taylor cone, and thus the formation of particles that are suitable candidates for drug and vaccine delivery applications.

Acknowledgements

Financial support for TS and AP, from the Clarkson University Honors Program, is greatly appreciated.

References

[1] Pillai, C. K. S., Paul, W., & Sharma, C. P. (2009). Chitin and chitosan polymers: Chemistry, solubility and fiber formation. Progress in Polymer Science, 34(7), 641-678. doi: 10.1016/j.progpolymsci.2009.04.001

[2] Myrick, J. M., Vendra, V. K. & Krishnan, S. (2014). Self-assembled polysaccharide nanostructures for controlled-release applications. Nanotechnology Reviews, 3(4), 319-346. doi: 10.1515/ntrev-2012-0050

[3] Bock, N., Woodruff, M., Hutmacher, D., & Dargaville, T. (2011). Electrospraying, a reproducible method for production of polymeric microspheres for biomedical applications. Polymers, 3, 131-149. doi: 10.3390/polym3010131

Manufacturing method of black ginseng prepared by repeated steaming and drying Process using loess mortar balls and its biological characteristics

Woo Sik Choi, Shigeki Toyama, KAPPIE, Busan, South Korea
Byung Sik Woo, Yanggihaek Health Function Manufacture Co., Geumsan, South Korea

Abstract

First of all, this paper explains the inherent manufacturing process of black ginseng, that is, put fresh ginseng in a kettle in which loess mortar ball is immersed, boil it, steam it, then dry. Repeat this process several times to prepare black bean ginseng. Next, it has comprehensively examined that the biological properties of this ginseng, basic and application experiments, literature, experiences of drinkers, etc., and the application to the natural healing efficacy as a kind of healthy function foods.

1. Introduction

In general, ginseng has already been used as a kind of healthy functional foods for thousands of years in Korea. Its efficacy and versatility are belonged to traditional Chinese herbal medicines. There is also a record of Korean red ginseng in Goryeo Gyeongsang of Xu Yong published about 1000 years ago in era of Song, China. It has been known to be a traditional herbal medicine processed and manufactured according to the traditional medicine technical process. In many cases, ginseng is used as it is or steamed sometimes. As the pharmacological action and clinical effect of ginseng, it has been reported such as central nervous system, excitement action, protein and nucleic acid biosynthesis promoting action, arteriosclerosis prevention, hypoglycemic effect, anti-fatigue, and anti-stress action etc. Recently, the effect of inhibiting the growth of AIDS virus and improving anti-dioxin and sexual function has been reported [1].

Ginseng is classified into fresh ginseng, white ginseng, taegeuk ginseng, and red ginseng according to processing methods. Red ginseng refers to dried ginseng by steaming or other methods. [2] During the steaming process, some components in the red ginseng are chemically converted and resulted in new physiologically active ingredients which are not in the fresh ginseng or white ginseng. Representative of these components are prosapogenin glycoside components including ginsenoside G-Rg3. [3-5]

From the early 2000s, a processing method of fresh ginseng to be repeatedly steaming and drying was developed based on the old traditional principle of so-called 'nine times steamed and dried process' at Geumsan area, Korea. The ginseng is called black ginseng because the color of fresh ginseng is changed to black color during repeated process of steaming and drying. [6,7]

Therefore, the black ginseng is significantly different from existing red ginseng in its appearance and biological characteristics. It has been found from researches that physiologically active ingredients, such as G-Rg3, which are present only in trace amounts in red ginseng, are significantly increased in black ginseng. [7-9] As a result, the amendment of the Ginseng Industrial Act is made in 2012 and it has become possible to distribute ginseng as a new item of ginseng products in addition to existing ginseng, white ginseng, taegeuk ginseng, and red ginseng.

In the case of South Korea, the representative products of functional ginseng are 'Seonsamjeong' developed in 2001. Palmerton Company of Boehringer Ingelheim, Switzerland has standardized human ginseng as a functional food and has achieved results in the world market, and it is used in China for anti-cancer treatment under the trade name "Shenyi" There [10,11].

Looking at the patents of the method of manufacturing black bean ginseng in Korea, the manufacturing method maximizing the content of crude saponin and ginsenoside through repeated steaming steps, the method of manufacturing by the existing method of nine times steaming and drying, the Korean classic production method with hydrolysis of starch in ginseng to increase ginsenoside contents to 1.00 to 7.00 mg/g and increase other contents have been published. [12].

As can be seen from the above research papers and patents, there are still many unknown parts about black ginseng, and expectations for the future of the basic and applied research are great [11].

In this study, the healing effects of black ginseng prepared with loess mortar balls immersed in the boiling kettle were investigated experimentally based on consideration for the healing experienced reports of users and literatures.

2. Experimental method

(1) Composition of loess mortar:

The loess mortar used in this study is composed of loess ocher 75 kg, mica 5 kg, mexomusoku 5 kg, vermiculite 5 kg, illite 4 kg, smacktite 3 kg, jadel powder 1 kg, germanite 1 kg and pojoran 1 kg.

South Korean loess which is the main ingredient of this composition has been traditionally used to make houses and earthenware etc. from old days, it gives warmth to the cold weather at night and it gives also coolness to the sunny day. This is said to be 'the energy of the earth'. In recent years the loess is gradually and easily used for daily necessities together with human hope. The healing efficacy of loess has been scientifically studied for pursuing a healthy life.

Fig. 1. A photograph of loess mortar balls

(2) Production process of loess mortar ball:

In the South Korean patent, the composition of loess mortar ball and the method for producing a molded item are the same as follows. [13] 75 kg of loess, 5 kg of mica, 5 kg of mexomusuku, 5 kg of vermiculite, 4 kg of illite, 3 kg of Smacktite, 1 kg of jadel powder, 1 kg of germanite, and 1 kg of pojolan are put in a ball mill and then pulverized to 200 to 350 mesh, sieved, 15 kg of water are added, and stirred to produce loess mortar. After kneading and molding the loess mortar, it is put in a heating furnace and calcined at 700 to 900°C for 4 hours.

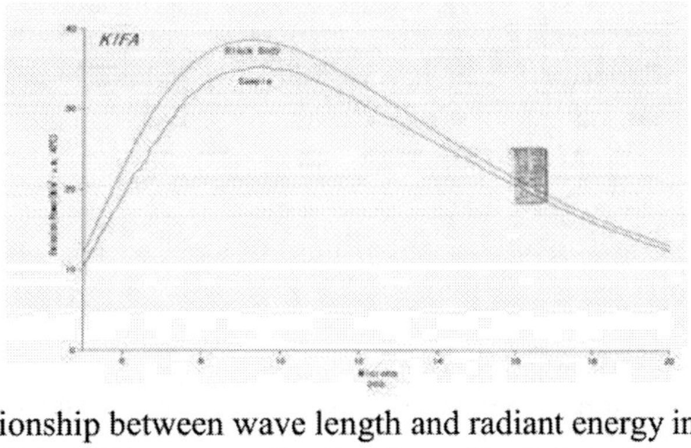

Fig. 2. The relationship between wave length and radiant energy intensity of loess mortal ball at 40°C measured by the Korea Far Infrared Association.

Furthermore, after the glaze treatment, calcining at 1200 to 1300°C for 12 to 15 hours is a main manufacturing step of preparing a loess mortar composition ball. A photograph of the prepared loess mortar balls is shown in Fig. 1.

The far infrared emissivity and radiant energy intensity of the loess mortar composition ball at 40°C measured by the Korea Far Infrared Association were 0.921 and 371 W/(($m^2 \cdot \mu m$) 40°C.), respectively as shown in Fig. 2. [14].

(3) Manufacturing method of ginseng:

After washing fresh ginseng of 3 years to 7 years, herein 5 years roots cleanly, 750 g of ginseng is brought to 3 liters of water and 3 kg of loess mortar ball is immersed in water in a pottery pot at the temperature of 60°C to 100°C. After heating for 2 to 3 hours, the processed ginseng is taken out. After aging for 24 hours at room temperature, firstly dry for 48 hours at a place where sunshine is good. After first steamed of the ginseng by heated steam with the loess mortar balls immersed in water for 2 hours, it is dried for 48 hours at sunny place. Furthermore, after second steamed of the ginseng by heated steam with the loess mortar balls immersed in water for 2 hours, and then it is dried for 48 hours at sunny place. And once again after third steamed of the ginseng by heating steam with the loess mortar balls immersed in water for 2 hours, and then it is dried for 48 hours. At last, a black ginseng is prepared as shown in Figures. A schematic diagram of the above manufacturing process is shown in Scheme 1 based on the application to patent [15].

Scheme 1. Overall manufacturing process of black ginseng and examination items of necessary mechanical specifications (on consideration)

Basic specifications:
- Discharge amount is 1 Batch water ginseng 1000 bowls (750 g / bowl × 1000 bowls = 750 kg)
- Box-type equipment (keeping the temperature distribution uniform during circulation of cold air or hot air of steam).
- Consider contact with eccentric nuclear material at the bottom, top, wall, etc. of the box.
- The inner shelf is considered to be a drawer type (the size and shape of the plate are an important key point on design).

A. Water ginseng cleaning process, for example, a bubble washing machine

B. Details of manufacturing process (steaming, steaming, drying process)

No. Process	**Main operating conditions and required time**
1. Boiling process	Boil water at 80 to 100°C and boil fresh ginseng in hot water for 0.5~1 hr.
2. Equipment of aging fermentation	Used as it is as aging fermentation room, temperature is 35~40°C for 48 hr.
3. Low-temperature drying	Drying with hot air of low-temperature in aged fermentation state at 40 to 60°C for 24 hr
4. Steam injection	Contents of # 3 are steam jet steam at 60 ~ 90°C as they are for 2-3 hours
5. Second drying	Place the contents of (4) as it is and dry with warm air at 40~ 60°C for 24 hr.
6. Second injection	Contents of No. 5 are steam 60 to 90°C. injection steam, after 2 to 3 hr, hot air drying, for 24 hr

7. Repeat process 5 and 6 three times.

C. Black Grain powder packaging (capsule filling) process

1. Milling process,
2. Granulation process,
3. Packaging (capsule filling) process

* Specially, the discussion of required hardware is desired for the whole and each process.

(4) Quantitative analysis test of indicator components in black ginseng:

The quantitative analysis test of indicator components in the prepared black ginseng samples was carried out at Korean Geumsan International Ginseng Research Institute. Using the Agilent 1260 HPLC System (Agilent Co. Ltd., USA) of the laboratory, the representative indicator components in black ginseng was mainly analyzed by the routine of already established analysis method. [1, 16].

(5) User's experiential case report:

Even though it is an experience case report, there is not a special form, but user's experienced healing case such as the motives of drinking, changes in the symptoms of the body before and after drinking, and overall impressions, etc. are reported in letters and telephone, and the reports are summarized for various purposes. [17].

3. Results and discussion

(1). Photos. observation of black-and-white ginseng during manufacturing processes:

Photographs taken for each process are shown in Fig. 3-1. From the figure, you can see the change in size and color of the body.

Raw fresh ginseng | First steamed and dried ginseng | Second steamed and dried ginseng | Final comparison of fresh and black ginseng

Fig. 3-1. Photographs of changes in size and color of fresh ginseng according to 3 steps of process progress.

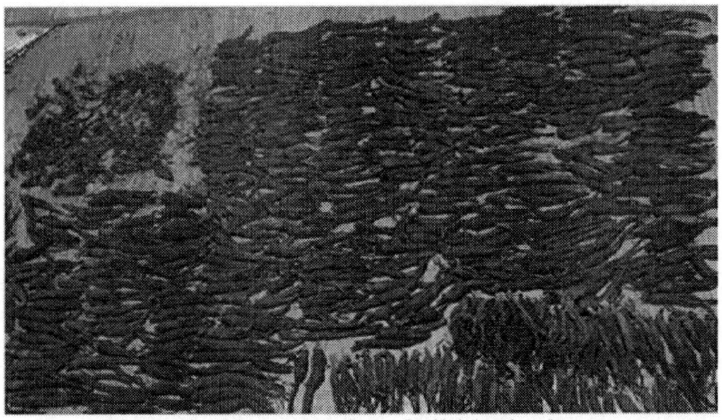

Fig. 3-2. Photographs of all black ginseng after the final manufacturing process of one batch preparation

Photos. of the final products of black ginseng after the final manufacturing process is shown in Fig. 3-2.

3-2. It turns out that dark brown black color of black ginseng becomes uniformly.

The carbohydrate in fresh ginseng is decomposed into 5-hydroxymethyl-2-furaldehyde (5-HMF) when the fresh ginseng is repeatedly steamed and dried. Next, the compound 5-HMF binds with amino acids and proteins to produce melanoidins so-called the final product of the Maillard reaction, [18] The color of the ginseng becomes dark brown. This phenomenon occurs also during the steamed process of other traditional medicinal herbs. [19] 5-HMF was not detected in white ginseng, but it increased gradually with increasing number of repeating steamed and dried process. [20]

(2) Test results for the quantitative analysis of ginsenoide species in black ginseng

Table 2 summarizes the test results for the quantitative analysis of ginsenoide species in black ginseng. The twice tests were carried out by Korean International Ginseng Herbal Medicine Research Institute and the test conditions are as followings: Items, criteria, results, judgment of test, and remark. [21].

FIG. 3-3 shows the chemical structures of ginsenosides species Rg 1, Rb 1, and Rg 3(S). [1,22]

Table 2. Test results for quantitative analysis of ginsenoide species in black ginseng.

- **Test Item**:	The sum of ginsenosides Rg 1, Rb 1, and Rg 3(S)
- **Test criteria**:	No specific criteria as general analysis method
- **Result**	Result of first test was 4.47 mg/g and one of second was 12.10 mg/g
- **Judgment**:	Results of the analysis data were confirmed as described.
- **Remarks**:	First run: Rg 1: 1.49 mg/g, Rb 1: 2.85 mg/g, Rg 3(S): 0.13 mg/g Second run: Rg 1: 2.05 mg/g, Rb 1: 9.00 mg/g, Rg 3(S): 1.05 mg/g Increase ratio (= First run/Second run): Rg 1:1.36, Rb 1:3.16, and Rg 3(S): 8.07

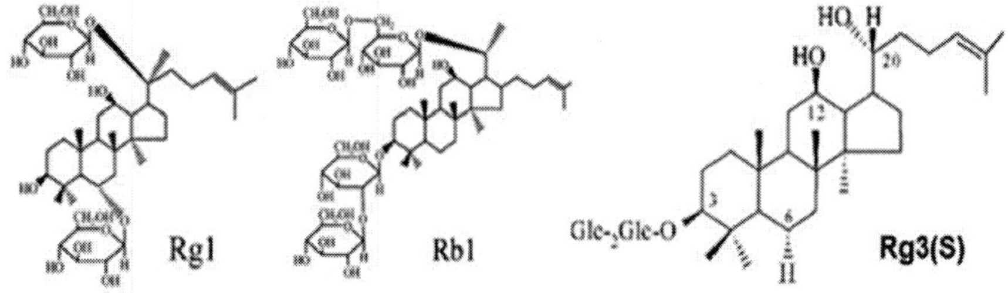

Fig. 3-3. Chemical structures of ginsenoide, three species, Rg 1, Rb 1, and Rg 3(S) [1, 22]

A developer of the black ginseng, who does not have the chemical analysis equipment of the prepared test samples had a meeting with the Korean International Ginseng Herbal Medicine Research Institute in Geomsan area and requested the quantitative analysis of three kinds of ginsenosides species Rgl, Rbl, and Rg3 (s) worthy of attention as some indicator components in black ginseng. Total contents of three ginsenoside species Rgl, Rb 1, and Rg3(S) in sample black ginseng was 4.47 mg/g in the first sample and 12.10 mg/g in the second sample. The three ginsenoside species Rgl, Rb 1, and Rg3(S) were hardly observed in the original fresh New ginsenosides species are produced during the production process of red ginseng. As its related mechanism, the hydrolysis, dehydration, and isomerization reaction at ginsenoside C-20 and the hydrolysis reaction at ginsenoside C-3 or C-6 is presumed to occur. [23] It was confirmed that the contents of ginsenoside prepared by this manufacturing method is high. It seems to be worthwhile to process it as a health function food, and a patent application to the Korean Patent Office is applied. [15].

Basic research for searching for optimal operating conditions to promote conversion to active ingredients is an important subject and urgent clinical trials through realization of industry-academic collaborative research are urgently required.

(3) Consideration on various healing effects from reports of the user's experienced case:

Among the numerous reports, three cases will be introduced briefly as follows. [17]

① Case of developers' experiences of 60's: For the restoration of fatigue level, the result of the continued observation for 3 months is as follows: the color of hair was progressing from white to black, the skin became soft, the foot athlete's foot was also disappeared. The promotion of male energies, the comfort of stomach, and the bowel movements became better.

② Case of male of 60's living in Busan: Skin was soft, body edema was gone, fatigue was resolved, gastric disorder was disappeared. Ten years ago, at a Busan hospital he was underwent surgery for colorectal cancer, and after that at a hospital with Busan due to cerebral hemorrhage, he was given hypoalgia. Bowel movements and the promotion of male energies were improved.

③ Case of male of 50's in Bucheon city, Gyeonggi-do Province: The skin got soft, the promotion of male energies was improved, the recovery of fatigue was smooth, and the stomach was comfort. After 3 month take, it was confirmed that the black ginseng was really good.

The healing effects of black ginseng have been confirmed by recent research literatures, [22-26] such as oxidative stress in human body [27], cancer [28], memory and intellectual performance, [29,30] diabetes, [31,32] blood pressure, [33,34] chronic inflammatory [35,36], and no significant toxicity. [37]

4. Conclusion

The method for producing black ginseng prepared by repeatedly heating, steaming and drying fresh ginseng with loess mortar balls immersed in water and the biological healing characteristics are examined based on consideration for reports of the user's experienced case and related literatures as healthy functional foods of the prepared black ginseng and the followings are summarized as the conclusions.

1. The total contents of ginsenosides in black ginseng were 12.10 mg/g.
2. Changes in the color of black ginseng observed with the naked eye were uniform dark brown.
3. Experienced user who has used the black ginseng for 3 months reported that constipation, thirst, tinnitus, dryness of the skin decreased and had a vigorous life to recover from fatigue etc..

In addition, it is known that this manufacturing method is worthy as a new processing step for health function foods.

References

1 Kii Yeul Nam, Jae Eul Choi, and Jong Dae Park, Transformation Techniques for the Large Scale Production of Ginsenoside Rg3, Korean J. Medicinal Crop Sci., 21(5), 401-414 (2013).

2. Jae Eul Choi, Ki Yeul Nam, Xiangguo Li, Bok Young Kim, Hang Sik Cho, and Kuang Bo Hwang, Changes of Chemical Compositions and Ginsenoside Contents of Different Root Parts of Ginsengs with Processing Method, Korean J. Medicinal Crop Sci., 18(2), 118-125 (2010).

3. Ki Yeul Nam, Yeong Su Kim, Mi Yae Shon, and Jong Dae Park, Recent Advances in Studies on Chemical Constituents and Biological Activities of Korean Black Ginseng (Panax ginseng CA Meyer), Kor., J .Pharmacogn., 46(3),173~188 (2015).

4. Jo ,E.J., Kang, S.J.and Kim ,A.J., Effects of steam- and dry processing temperatures on the benzopyrene contents of black and red ginseng, J. Korean Soc., Food Sci. Nutr., 22:199-204 (2000).

5. Han, S.T.,Whang, W.K.,Kim ,I.H.,Yang, B.W., Cho, S.H. and Ko, S.K., Analysis of ginsenosides of black ginseng, Yakhak Hoeji, 49:490-494 (2005).

6. Hong, H.D., Kim, Y.C., Rho, J.H. ,Kim, K.T. and Lee, Y.C., Changes on physiological properties of Panax ginseng C.A .Meyer during repeated steaming process, J. Ginseng Res.,31, 221-229 (2007).

7. Jee Hyun Lee, Gui Nan Shen, Eui Keom Kim, Hyun Joong Shin, Chang Seon Myung, Han Jin Oh, Dong Hee Kim, Seong Soo Roh, Won Cho, Young Bae Seo, Yong Jin Park, Cheol Woo Kang, Gyu Yong Song, Preparation of Black Ginseng and its antitumor activity, Korean J. Oriental Physiology & Pathology, 20(4):951-956, (2006).

8. Kitagawa I, Taniyama T, Shibuya H, Noda T and Yoshikawa M., Chemical studies on crude drug processing. V, On the constituents of ginseng radix rubra(2):Comparison of the constituents of white ginseng and red ginseng prepared from the same Panax ginseng root, Yakugaku Zasshi, 107, 495-505 (1987).

9. Kim ND, Kim EM, Kang KW, Cho MK, Choi SY, and Kim SG, Ginsenoside Rg3 inhibits phenylephrine-induced vascular contraction through induction of nitric oxide synthase, British Journal of Pharmacology, 140, 661-670 (2003).

10. Nam KY, The Comparative understanding between red ginseng and white ginsengs, processed ginsengs (Panax ginseng CA Meyer), Journal of Ginseng Research, 29, 1-18 (2005).
11. Keum,YS, Park, KK, Lee, JM, Chun, KS, Park, JH, Lee, SK, Kwon, H., Surh, YJ, Antioxidant and anti-tumor promoting activities of the methanol extract of heat-processed ginseng, Cancer Lett,150(1), 41-48 (2000).
12. http://www.kipris.or.kr, Korea Intellectual Property Rights Information Service (2013).
13. Functional ceramics positive nuclei products and processes for emulsifying,curing and emitting far infrared rays, Patent Publication No. 10-2004- 0104287 (2004).
14. Certificate of FIR test analysis for loess mortal ball, KICM (2015).
15. Woo, Byung Sik, Choi, Woo Sik, Jeong, Jang Chun, Method for manufacturing black ginseng for enhancing immunity, Application No. 10-2017- 0098763,in Patent Application (2017).
16. LabSolutions Analysis Report measured by Korean International Ginseng Herbal Medicine Research Institute (2017).
17. Report on the experience of black ginseng, health function positive nuclear manufacturing company (2017).
18. Maillard, L.C, Formation of melanoidins in a methodical way. Compt. Rend. 154:66. (1912).
19. Sun, B.S., Pan You, F.U. and Sung, C.K., Repetitious steaming-induced chemical transformations and global quality of black ginseng derived from Panax ginseng by HPLC ESI-MS/MSn based chemical profiling approach. Biotechnol. Bioprocess Eng. 16: 956-965. (2011).
20. Yang, S.J., Woo, K.S., Yoo, J.S., Kang, T.S. and Noh, Y.H., Change of Korean ginseng components with high temperature and pressure treatment. Korean J. Food Sci. Technol. 38: 521-525. (2006).
21. Results of the first and 2nd test analysis of Geumsan International Ginseng Herbal Research Institute (2017).
22. Ki Yeul Nam, Yeong Su Kim, Mi Yae Shon, and Jong Dae Park, Recent Advances in Studies on Chemical Constituents and Biological Activities of Korean Black Ginseng (Panax ginseng CA Meyer), Kor. J. Pharmacogn., 46(3),173~188 (2015).
23. Kikuji Yamashita, The Effects of the Far-Infrared Ray (FIR) Energy Radiation on Living Body, "Blood Cell- An Overview of Studies in Hematology", book edited by Terry E. Moschandreou, (2012),
24. K. Yamashita, Study of Life Science Based on Anatomy: True Characteristics and Effects of Far Infrared Ray (FIR) Energy Radiation, Journal of oral health and biosciences, 2792), 97-105, (2015).
25. Huh, J.: "Donguibogam translated by Hun Young Cho", pp. 2753-2758, Publisher Yeogang (2005).
26. Yang, W.Z., Hu, Y., Wu, W.Y., Ye, M. and Guo, D.A., Saponins in the genus Panax L. (Araliaceae): A systematic review of their chemical diversity, Phytochemistry 106: 7-24. (2014).
27. Giustarini, D., Dalle-Donne, I., Tsikas, D. and Rossi, R., Oxidative stress and human diseases: origin, link, measurement, mechanisms, and biomarkers. Crit. Rev. Clin. Lab. Sci. 46: 241-281. (2009).
28. Kim, E. K., Lee, J. H., Cho, S. H., Shen, G. N., Jin, L. G., Myung, C. S., Oh, H. J., Kim, D. H., Yun, J. D., Roh, S. S., Park, Y. J., Seo, Y. B. and Song, G. Y., Preparation of black Panax ginseng by new methods and its antitumor activity. Kor. J. Herbology 23: 85-92, (2008).
29. Nishijo, H., Uwano, T., Zhong Y. M. and Ono, T., Proof of the mysterious efficacy of ginseng: basic and clinical trials: effects of red ginseng on learning and memory deficits in an animal model of amnesia. J. Pharmacol. Sci. 95: 145-152 (2004).
30. Zhong, Y. M., Nishijo, H., Uwano, T., Tamura, R., Kawanishi, K. and Ono, T., Red ginseng ameliorated place navigation deficits in young rats with hippocampal lesions and aged rats. Physiol Behav. 69: 511-525. (2000).
31. Yuan, H. D., Kim, J. T., Kim, S. H. and Chung, S. H., Ginseng and Diabetes: The Evidences from In Vitro, Animal and Human Studies. J. Ginseng Res. 36: 27-39. (2012).
32. Vuksan, V., Sung, M. K., Sievenpiper, J. L., Stavro, P. M., Jenkins, A. L., Di Buono, M., Lee, K. S., Leiter, L. A., Nam, Y. and Arnason, J. T., Korean red ginseng (Panax ginseng) improves glucose and insulin regulation in well-controlled, type 2 diabetes: results of a randomized, double-blind, placebo-controlled study of efficacy and safety. Nutr. Metab. Cardiovasc. Dis. 18: 46-56. (2008)
33. Gillis, C. N., Panax ginseng pharmacology: A nitric oxide link, Biochem. Pharmacol. 54: 1-8. (1997)
34. Jeon, B. H., Kim, C. S., Kim, H. S., Park, J. B., Nam, K. Y. and Chang, S. J., Effect of Korean red ginseng on blood pressure and nitric oxide production. Acta Pharmacol. Sin. 21: 1095-1100. (2000)
35. Bae, E.A., Joo, H.M., Shin, Y. W. and Kim, D. H., Inhibitory effects of Korean red ginseng and its genuine constituents ginsenosides Rg3, Rf, and Rh2 in mouse passive cultaneous anaphylaxis reaction and contact dermatitis models. Biol. Pharm. Bull. 29: 1862-1867 (2006).
36. Trinh, H. T., Shin, Y. W., Han, S. J., Han, M. J. and Kim, D. H., Evaluation of antipruritic effects of red ginseng and its ingredients in mice. Planta Med. 74: 210-214. (2008).
37. Chani, P.C. and Fu, P.P., Toxicity of Panax ginseng - An herbal medicine and dietary supplement. J. Food Drug Anal. 15: 416-427. (2007).

509046 Preparation of Glucan Microparticles with Curcumin for Treatment of Idiopatic Inflammation Diseases

Abstract:

One of the novel methods of pharmaceutical formulation is the loading of the drug into biocompatible microparticles, preferably extracted from natural sources. This approach offers several potential advantages, *e.g.* increased bioavailability (the ratio of the administered drug, which actually enters the bloodstream) compared to conventional drug delivery methods, enabling the option of oral administration even for parenteral drugs. Multiple principles of the loading process are possible – drug encapsulation, adsorption onto the particle surface or precipitation inside the pores a hollow particle. The presented work focuses on precipitation of a drug from an organic solvent inside glucan particles. Glucan microparticles are ellipsoid porous shells, which are obtained from cell walls of *Saccharomyces cerevisiae* (baker´s yeast). The particle size is 2-5 micrometers. Cell walls of glucan particles are mainly composed of β-1,3-glucans. In order to prevent inflammatory reaction in organism, the yeast organelles are removed by a series of alkaline and acidic extraction steps. Subsequently, present lipopolysacharides are extracted using different organic solvents.
Lymphatic cells are mainly present in inflammation area in organism where they are responsible for progress of inflammation. Glucan particles (GPs) interact with lymphatic cell´s receptors on their surface and after that β-1,3-glucans are phagocytosed by the lymphatic cells. Therefore, GPs represent an effective oral delivery system for transporting of biological active substances. Curcumin is an anti-inflammatory flavonoid, which occurs in plants of family *Zingiberacaea.* In our project, curcumin was delivered to an inflammation in organism using GPs. Loading of curcumin into GPs was enabled via precipitation of curcumin from organic solvents. Curcumin dissolved in organic solvents was added to the GPs and through pores penetrated into cavity of glucan shell. Next, the solvent was removed and the curcumin was precipitated in glucan microparticles. GPs with the flavonoid inside were observed using confocal microscopy and SEM (scanning electron microscopy). The achieved content of curcumin was successfully analyzed by other methods, which can not be mentioned directly due to article in preparation. Furthermore, the curcumin-loaded GPs were tested *in vitro* on macrophage like cells and *in vivo* on laboratory mice in order to evaluate their effectivity. The obtained results show that these composites are promising materials for biomedical applications such as efficient treatment of inflammatory diseases (e.g. idiopatic inflammatory bowel disease, rheumatoic arthritis).

508600 Long-Term Dissolution Method for Intramuscular Suspension

Abstract:

Sustained release (SR) injectable suspensions – also called "nanoparticle depot systems" – are already used in clinical practice because they have a number of advantages: lower frequency of administration, blood levels remain within therapeutic range, greater effectiveness and reduced side effects, as well as improved patient compliance. Such suspensions may be administered in the form of a low soluble drug, but the release time of such medication is dependent on the particle size and the specific surface area. In drug development, it is very important to have reliable method how to evaluate drug dissolution profile in *in vitro* conditions before clinical trials on animals and humans. In the case of low soluble drug, common dissolution tests are not applicable because of drug insolubility. Therefore, we have developed a new *in vitro* dissolution method to evaluate the dissolution profile of such nanoparticle medication. Moreover, this method is also useful for the development of generic drugs, in order to compare their dissolution profile to those of the reference. Enzymatic dissolution tests of our drug suspensions were conducted with various PSD by the enzyme serine esterase in a physiological solution at 37 C. We prepared those suspensions of various PSD by wet nano milling in a ball mill containing ZrO_2 beads, based on previous parametric study where we investigated the impact of variable process parameters such as time, ball diameter, fill level and agitation rate. For each sample taken at various time intervals, the increase of API concentration and reduction of PSD of the suspension were evaluated using HPLC and laser diffraction, respectively. Based on the results of this work, the rate of kinetic release was found to correlate directly with an increase in the specific surface area of the particles in the suspensions. The proposed in vitro method is used to predict the pharmacokinetics of the medication under in vivo conditions. Thus, it can serve as a cheaper and faster alternative for bioequivalence studies (BES) on animals used in generic drug development. However, further validation studies will be required to relate our method quantitatively to real *in vivo* conditions.

508013 High Drug-Loaded Surfactant-Free Nanocomposite Microparticles for Enhanced Dissolution of Poorly Soluble Drugs

Abstract:

About 80% of newly discovered drug molecules pose some poor water-solubility related issues, which makes them difficult to formulate into efficacious medications. Dissolution of a poorly soluble drug can be improved by breaking the drug crystals down to nano-size in the form of a nanosuspension and subsequently drying it into nanocomposite microparticles (nanocomposites). A major challenge in this field is that depending on the dispersants used, drug nanoparticles may aggregate, and they may not be recovered fast during dissolution from the nanocomposites. The aim of this study is to investigate the impact of various classes of dispersants on the dissolution of drug nanoparticles from nanocomposites and elucidate the mechanisms for enhanced dissolution. Precursor suspensions of itraconazole (ITZ), a model poorly water-soluble drug, in the presence of various dispersants were prepared via wet stirred media milling and spray dried to form nanocomposites. Hydroxypropyl cellulose (HPC, water-soluble polymer) alone, sodium dodecyl sulfate (SDS, anionic surfactant) alone, and their combination were used as baseline stabilizers/dispersants. Mannitol and Sucrose were used as typical water-soluble dispersants in addition to HPC. As a major novelty of this investigation, swellable dispersants such as wet-milled superdisintegrants were used in the nanocomposites. Specifically, three superdisintegrants most commonly used in pharmaceutical industry, i.e., sodium starch glycolate (SSG), crospovidone (CP), and croscarmellose sodium (CCS) were wet-co-milled along with ITZ. Laser diffraction, viscometry, SEM, XRD, DSC, UV spectroscopy, wettability testing, and dissolution testing were used for characterization. Our experimental results indicate that HPC–SDS combination or higher HPC concentration allowed for faster drug dissolution due to reduction of drug nanoparticle aggregation and enhanced wettability. Wet co-milled superdisintegrants were much more effective than Mannitol and Sucrose in dissolution enhancement of ITZ. With judicious selection of superdisintegrant concentration, their performance matched that of HPC–SDS combination, suggesting that the use of wet-milled superdintegrants obviates the need for surfactants that could be toxic, especially at high concentrations for various pharmaceutical applications. The impact of the wet-milled superdisintegrants positively correlated with their swelling capacity (SSG > CCS > CP), signifying the dominance of a swelling-induced erosion/disintegration mechanism during drug dissolution from the nanocomposites with wet-milled superdisintegrants. This study overall demonstrates that fast-dissolving, high drug-loaded, surfactant-free nanocomposites could be prepared with either HPC co-milled superdisintegrants or optimal HPC concentration via spray drying. Since many patients find surfactants to be irritating in particular for orally dispersing dosage forms, such surfactant-free drug composites are expected to alleviate that aspect of patient compliance issues.

507960 Convective Drying Kinetics of Polymer Strip Films

Abstract:

Polymer strip films are flexible platforms for oral delivery of drugs. They have shown superior performance compared to the traditional solid dosage forms[1]. These films are extensively used as a delivery form for water-soluble drugs but they can be promising candidates for the delivery of poorly water-soluble drugs, which attracted attention in recent years[1]. Drying of strip films is an important part of the manufacturing process. Drying process has significant effect on the final morphology of the strip films and consequently on their quality and performance. In this research, a comprehensive kinetic study is done based on design of experiments (DOE) and integral method of analysis to obtain models for convective drying of polymer strip films loaded with poorly water-soluble drug particles, at different process conditions. The major novelty of this work is in examining the drying kinetics and corresponding the model development that incorporate the impact of drug particles. These models can be used to analyze the effect of process parameters and optimally design the drying process to produce strip films with the desired moisture content and thickness. In the experimental part of this study, a polymer solution of hydroxypropyl methylcellulose (HPMC) E-15 and glycerin mixed with drug suspension (fenofibrate) was used in a doctor blade to cast films on a plastic substrate at a certain thickness. Drying data was collected using a batch set-up in which the air velocity and temperature were controlled by an air blower with a heating coil. The plastic substrate was placed on a heated metal plate with a fixed temperature. The plate was positioned on a digital scale connected to a computer system and the weight of the film was recorded every 2 s. The air temperature and velocity in the chamber were monitored throughout the process, using thermocouples and an anemometer to ensure steady-state operation. Data was collected based on a full factorial design (DOE) with center point for air temperature (40, 55 and 70 ÌŠC) and velocity (0.2, 0.6 and 1 m/s). A drying mechanism consisting of a constant-rate and a falling-rate period was assumed and based on that a rate model was used in the nonlinear regression to fit the experimental data. The model shows good agreement with the experimental data for the moisture content. At lower air temperature and velocity, shorter constant-rate period and lower drying rate are observed. By assuming moisture transport in the gas phase above the film as the limiting transport mechanism, the constant drying rate was also calculated using principles of heat transfer and available correlations for heat transfer coefficient. The result shows a difference between the estimated and calculated constant drying rates. Overall, this study presents a methodology for estimating the drying rates of wet polymeric strip films in a convective batch drying set-up, which is expected to help facilitate development of robust drying processes. Our study also demonstrates the need for a more detailed transport model to simulate the drying process, which is currently in progress.

1. Krull, Scott M., et al. "Polymer strip films as a robust, surfactant-free platform for delivery of BCS Class II drug nanoparticles." *International journal of pharmaceutics* 489.1 (2015): 45-57.

507884 The Effects of Mixing Processes on Critical Quality Attributes of Strip Films

Abstract:

This work is part of an overall objective of being able to create patient compliant drug loaded thin polymeric films that are uniform in drug content or where the amount of drug varies by a minuscule amount and has negligible effects to the drug dosage. Three mixing processes, impeller, planetary, and high-intensity vibratory mixer (HIVM), are investigated for preparation of polymer-drug slurry based film precursors. The study examines key process parameters and key formulation parameters and their effect on critical quality attributes of strip film. Since each mixing process has different mixing efficiencies, a proper design of experiment (DOE) study is conducted to better understand the processes. Key DOE variables were process intensity, process time, and the mixing device. The drug used was fenofibrate (FNB) and the drug loading used to determine these three parameters was 20% in the end product. It was found that of the three mixing processes, HIVM may be the poorest choice because it tends to give sporadic results with different parameters for the uniformness of the precursor, which leads to uneven drug distribution in the dried films. The film product drug content uniformity, measured using relative standard deviation (RSD) for intermediate and end-product, was better correlated for the planetary mixer, and somewhat correlated for the impeller mixer. Based on the DOE outcomes, the mixing parameter with the lowest RSD for each mixer type was selected to be used in creating homogeneous precursors with low drug loadings; ranging from 6% to 0.6% FNB in the dried film. The outcomes indicate that the planetary mixer with its higher-intensity mixing capability allows for a more uniform product and in a shorter processing time. It is worth noting that the RSD of films is based on 1/10th the dose in this study. Thus, the overall results suggest that both the impeller and planetary mixers are successful in preparing film precursors for product with acceptable CQAs even at 0.6 % drug loading.

508731 Comprehensive Study of Usage of Oil Liquid Marbles in Pharma Industry

Abstract:

Many water poorly soluble active pharmaceutical ingredients (APIs) have been developed that tend to lower bioavailability. Recently, many papers are focused on a preparation of a new dosage form to overcome this problem. Liquid oil marble (LOM) is a droplet of oil encapsulated by solid particles. Creation of hard gelatin capsules filled by the LOMs containing the API was suggested in this work. The API was dissolved in oil, then small droplets of mixture were poured onto the solid particles and LOMs were created by shaking. Here we present parametric study of different combinations of solid particles (30 types) and oils (3 types). We characterize critical parameters of LOMs such as time stability, mass fraction of oil, API and mechanical stability. After hard gelatin capsules were filled by the LOMs with API and dissolution curves were measured. We compared the dissolution curves with commercial products. LOMs were successfully prepared from all combinations but LOMs with volume of 10μl were stable for longer period. The oil mass fractions were generally higher in combinations of methylcellulose with different oils. The highest load before failure occurred in case of methylcellulose with different oils. Our results suggest that the LOMs could be effective dosage form for poorly water soluble APIs.

509666 Impermeable Metal Nano-Capsules for Drug Delivery without Side Effects

Abstract:

Impermeable metal nano-capsules for drug delivery without side effects

Cancer therapeutics have dramatic side effects on healthy tissues. A prominent research area focuses on encapsulating cytotoxic drugs for targeted delivery to cancer tissues and for reduction of off-tissue side-effects. However, significant challenges remain for encapsulated clinical drugs (e.g. liposomal doxorubicin):

1. Drug encapsulation remains very expensive.

2. Drug loading is low and the manufacturing process is inefficient

3. Once encapsulated drug leaching over time is often high (especially on dilution)

4. Typically release specifically within tumours is not achieved

Recently, we have demonstrated the efficient manufacture of impermeable metal-shell/liquid core microcapsules (Hitchcock 'Long-term Retention of Small, Volatile Molecular Species within Metallic Microcapsules', ACS Appl. Mater. Interfaces, 2015, 7, 14808) that permit localised delivery of active (pharmaceutical) ingredient high doses by triggering release with ultrasound at the target location. This delivery method has the potential to be superior to all previously developed encapsulation strategies because it would address all of the above challenges simultaneously:

1. Capsules can easily be manufactured at industrial scale,

2. High drug content can be achieved within capsule cores,

3. No drug leaching occurs, thus preventing any side-effects prior to release activation via ultrasound treatment,

4. Complete release of high drug concentrations can be achieved in cancer-affected areas only.

Recent Relevant Publications:

- 4 recent patents – James Hitchcock, Coated microcapsules 20160168509, 20160168511, 20160177221 and 20160168510 publication dates 16/6/16
- The effect of surfactant chain length on the morphology of poly(methyl methacrylate) microcapsules for fragrance oil encapsulation Alison Tasker*, James Hitchcock, Ling He, Elaine Baxter, Simon Biggs and Olivier J. Cayre (Journal of colloid and interface science Volume 484, 15 December 2016, Pages 10–16)
- Long-Term Retention of Small, Volatile Molecular Species within Metallic Microcapsules J. P. Hitchcock*, A. L. Tasker, E. A. Baxter, S. Biggs, and O. J. Cayre, ACS Applied Materials & Interfaces, vol. 7, pp. 14808-14815, 2015/07/15 2015
- Adsorption of catalytic nanoparticles onto polymer substrates for controlled deposition of metal microcapsule shells J. P. Hitchcock*, A. L. Tasker, S. Biggs, and O. J. Cayre (submission to ACS Langmuir Feb, 2017)

507693 Investigation of Different Continuous Drying Routes for Pharmaceuticals

Abstract:

Batch processing was the state of the art production route for pharmaceuticals during the last decades. Although widely established, this way of manufacturing has some drawbacks such as higher footprint of the entire production site, complex logistics, limited possibilities for in-line process control and cost intensive scale-up. In order to increase robustness, flexibility and cost-effectiveness of manufacturing, academia, industry and regulators have been pushing the development of fully continuous lines. Continuous processing allows utilization of the same equipment for process development as well as for the final production process. In addition, robust product quality can be guaranteed by an in-process quality control. A full continuous pharmaceutical manufacturing facility may include API synthesis as well as manufacturing of the final dosage form. A broad range of continuous manufacturing equipment is already on the market for various unit operations, but there is still a gap to link primary and secondary manufacturing.

One of those challenging process steps between primary and secondary manufacturing is the continuous drying of APIâ €™s. During API synthesis and subsequent crystallization the desired particle properties, such as size and morphology, are tailored. However, during washing and drying these properties can be changed, due to attrition and agglomeration. Therefore, the challenge during drying is to minimize such effects.

This presentation will highlight results of drying experiments with two different drying technologies. APIs and excipients with a particle size below 100 Âµm were dried either based on extrusion or contact-convective drying technology. Variation in process settings was tested and successful configurations were found. Feed materials with different initial moisture content were processed to evaluate the drying capacity of the two dryer options. Measurements of particle size distribution, residual moisture content and flowability were conducted. Residual moisture of the dried products reached levels below 0.1%. No change in particle size distribution for the tested APIs was observed and therefore no agglomeration or attrition occurred during the drying procedure. In addition, powder flowability showed similar results prior and after drying.

The key to successful continuous drying of powders and matching specified product quality is a robust process design with consistent processability. This includes avoiding of dead zones, resulting in a narrow residence time distribution (RTD), amongst others. A narrow RTD leads to less mixing of particles with different moisture levels and results in a reduced tendency of agglomeration or attrition.

509524 A Systematic Study of Liquid and Solid Residence Time Distributions in a Dynamic Baffle Crystallizer

Abstract:

Crystallization has become a predominant technique in particle design technology and it is widely applied in the pharmaceutical industry. It directly dictates downstream processes and heavily influences overall drug properties such as bioavailability and dissolution rate. Traditionally crystallization has been carried out in batch mode where operating conditions vary with time, often resulting in inconsistent products.[1] Continuous crystallization, on the other hand, has been identified to improve reproducibility, reduce operating cost and simplify scale-up.[2] A well-studied system for continuous crystallization operation is stirred tank reactors (STRs) operated as Mixed-Suspension-Mixed-Product-Removal (MSMPR) crystallizers where an agitator is commonly used to provide mixing. There are challenges associated with stirred tank systems such as poor local mixing, low heat and mass transfer, and varying shear rate.[3] To address these challenges, a novel commercial dynamic baffle crystallizer (DBC) has been studied. It consists of a reactor tank and dynamic â€˜donutâ €™ shaped baffles to provide oscillatory flow resulting in more uniform mixing.[4] Oscillatory systems have been well studied for batch reactions and synthesis. Interest in the use of DBCs for crystallization processes has increased due to enhanced mass and heat transfer capabilities and reduced shear imposed on crystals.[5]

In the work proposed here, the DBC is operated as a MSMPR vessel. A systematic study of residence time distribution (RTD), essential for studying the hydrodynamics of any new vessel, is carried out. A statistically sound design of experiment (DOE) is performed to investigate the effect of frequency, amplitude and flow rate. FTIR probe, capable of correcting for signal interference caused by oscillations, is used to monitor the solute concentration during tracer experiments. This study will improve the understanding of oscillatory systems as an alternative crystallization vessel and the understanding of the impact of different mixing mechanisms on the final crystal size distribution.

1. Nagy ZK, Aamir E. Systematic design of supersaturation controlled crystallization processes for shaping the crystal size distribution using an analytical estimator. *Chem Eng Sci*. 2012;84:656-670. doi:10.1016/j.ces.2012.08.048.

2. Alvarez AJ, Singh A, Myerson AS. Crystallization of Cyclosporine in a Multistage Continuous MSMPR Crystallizer. *Cryst Growth Des*. 2011;11(10):4392-4400. doi:10.1021/cg200546g.

3. Kacker R, Regensburg SI, Kramer HJM. Residence Time Distribution of Dispersed Liquid and Solid Phase in a Continuous Oscillatory Flow Baffled Crystallizer. *Chem Eng J*. 2017;317:413-423. doi:10.1016/j.cej.2017.02.007.

4. Hewgill MR, Mackley MR, Pandit AB, Pannu SS. Enhancement of gas-liquid mass transfer using oscillatory flow in a baffled tube. *Chem Eng Sci*. 1993;48(4):799-809. doi:10.1016/0009-2509(93)80145-G.

5. Ni X, Mackley MR, Harvey AP, Stonestreet P, Baird MHI, Rama Rao NV. Mixing Through Oscillations and Pulsationsâ €"A Guide to Achieving Process Enhancements in the Chemical and Process Industries. *Chem Eng Res Des*. 2003;81(3):373-383. doi:10.1205/02638760360596928.

506480 The Axial Dispersion Performance of a Novel Oscillatory Flow Reactor with Liquid Solutions and Solids Suspensions - a Design of Experiments Approach

Abstract:

Introduction

The oscillatory flow reactor (OFR) has been widely investigated for the last two decades as a novel platform for achieving efficient and uniform mixing at average shear rates up to one order of magnitude lower than the standard stirred tank reactor (STR) [1]. The OFR consists of a tubular device containing evenly spaced baffles which are transversely assembled to a periodically oscillating flow. The mixing is achieved by the interaction of the fluid with the constrictions, which may be designed with different configurations according to the application [2].

The novel OFR presented in this work was designed with smooth periodic constrictions and differs from the typical integral baffle configuration in the cross-section format, which is rectangular rather than circular. This new feature was developed to improve the suspension and the flow of solid particles in continuous processing operations, such as crystallization. In order to characterize the mixing efficiency of this device, residence time distribution (RTD) experiments with liquid solutions and solid suspensions were conducted by monitoring the response of a pulse tracer for each system. A face-centered composite experiment was designed through a design of experiments (DoE) approach to assess the effects of different hydrodynamic conditions on the axial dispersion of liquids and solids.

Experimental Section

The OFR used in this work consists of a 3.3 m long Perspex® plate-like reactor, with an internal diameter of 8 mm in the straight section and 3.3 mm in the constriction, and a total volume of 101.5 mL. A continuous net flow of deionized water was injected into the reactor by a peristaltic pump and oscillated by a linear motor. At time zero, a pulse tracer of 2 mL was quickly injected at 25 cm downstream of the water inlet. A 0.4 g/L aqueous solution of indigo carmine was used as liquid tracer and a 40 % (m/m) suspension of PVC particles with a mean diameter of 160 m was used as solid tracer. The concentration of tracer was measured in the reactor outlet by a spectrometer at the wavelengths of 581 nm and 655 nm for the liquid and the solid tracers, respectively. The Peclet number was determined by fitting the experimental data to a plug flow with axial dispersion model [3].

A full factorial experiment was designed with Minitab® to assess the effect of three factors, the net flow rate, Q, the frequency of oscillation, f, and the amplitude of oscillation, x_0, on two response variables, the Peclet number of liquids, Pe_L, and the Peclet number of solids, Pe_S. Each factor was studied at two levels: 16 and 109 ml/min for Q, 3 and 6 Hz for f, and 3.9 and 6.5 mm (center-to-peak) for x_0. In coded units, the low level is defined as −1 and the high level is defined as +1. Each combination of factors was performed in triplicate. This design was augmented with centre and axial points and turned into a face-centered composite design in order to build a second order model for each response variable. Each term of this model is either linear, square or an interaction between factors and has an associated coefficient that delivers the best fit between the model and the experimental data.

Results and Discussion

A second-order model was developed in Minitab® so as to predict Pe_L and Pe_S with different levels of Q, f and x_0. The significance of each term of the model was determined by comparison of the p-value of each term to the significance level of 90 %. The p-value measures the evidence against the null hypothesis, which states that there is no association between the term and the response. A backward stepwise elimination technique was applied to select the significant terms, and to eliminate the insignificant terms from the model. All the linear terms, the square term of f, and the interactions terms Qx_0 and fx_0 were found significant for both response variables. The final regression models for Pe_L and Pe_S are described by the following equations, in coded units:

$$Pe_L = 182.0+151.98Q+66.84f-32.78x_0+92.5f^2-21.79Qx_0+20.41fx_0$$

$$Pe_S = 246.0+194.85Q+70.08f-57.78x_0+101.9f^2-53.16Qx_0+13.97fx_0$$

These models have coefficients of determination of 96.3 % and 97.3 %, respectively. Also, the p-values for the lack-of-fit were found higher than 0.10 for both models, so there is no evidence that the models do not fit the data with a confidence interval of 90 %.

The regression coefficients of both models have the same order of magnitude and the same signal, meaning that the direction and the strength of each factor is very similar for both Pe_L and Pe_S. Moreover, the positive effects of Q and f, and the negative effect of x0 are consistent with previous studies [3]. The contour plots also show similar trends for both models. Therefore, one can conclude that liquid solutions and solid suspensions have identical dispersion performances in this reactor.

Conclusions

A novel OFR with smooth periodic constrictions and a rectangular cross-section was developed. RTD experiments with liquid solutions and solid suspensions were performed to determine Pe_L and Pe_S in this device. A face-centered composite experiment was designed and a second-order model was built so as to predict Pe_L and Pe_S with different levels of Q, f and x_0.

The novel OFR used in this work was found to have similar axial dispersion performances with liquid solutions and solid suspensions. This is of paramount importance for multiphase systems such as crystallization where the migration of molecules between a supersaturated solution and crystalline particles takes place throughout the whole nucleation and growth steps.

The models obtained for both Pe_L and Pe_S might be further improved by the addition of other significant variables that were not considered in this study, namely the pressure drop. This might be object of future work.

Acknowledgments

This work was the result of the project: I) Project POCI-01-0145-FEDER-016816 (PTDC/QEQ-PRS/3787/2014) funded by the Project 9471 – Reforçar a Investigação, o Desenvolvimento Tecnológico e a Inovação (Project

Tecnologia I.P. (FCT); II) IF exploratory project [IF/01087/2014] funded by FCT; III) POCI-01-0145-FEDER-006939 (Laboratory for Process Engineering, Environment, Biotechnology and Energy – UID/EQU/00511/2013) funded by the European Regional Development Fund (ERDF), through COMPETE2020 - Programa Operacional

NORTE 01 0145 FEDER 000005 – LEPABE-2-ECO-INNOVATION, supported by North Portugal Regional Operational Programme (NORTE 2020), under the Portugal 2020 Partnership Agreement, through the European Regional Development Fund (ERDF). P. Cruz gratefully acknowledges doctoral scholarship [SFRH/BD/119391/2016] from FCT. A. Ferreira is an Investigador FCT.

References

[1] X. Ni, J. A. Cosgrove, A. D. Arnott, C. A. Greated, and R. H. Cumming, "On the measurement of strain rate in an oscillatory baffled column using particle image velocimetry," Chem. Eng. Sci., vol. 55, no. 16, pp. 3195–3208, 2000.

[2] J. R. McDonough, A. N. Phan, and A. P. Harvey, "Rapid process development using oscillatory baffled mesoreactors – A state-of-the-art review," Chem. Eng. J., vol. 265, pp. 110–121, 2015.

[3] M. Zheng and M. Mackley, "The axial dispersion performance of an oscillatory flow meso-reactor with relevance to continuous flow operation," Chem. Eng. Sci., vol. 63, pp. 1788–1799, 2008.

509212 CFD Modelling of Drying and Crystalline Transformation in a Co-Current Spray Dryer

Abstract:

Spray drying is widely applied in the pharmaceutical industry for the formulation of particulate materials such as solid dispersions, excipients, active pharmaceutical ingredient (API) as well as co-dried materials (API with excipient). It has a distinct advantage of converting pumpable feed into dry powder in a single unit operation. In spray drying, the atomised droplets typically ranging from 10 to 500 Î¼m in diameter comprise dissolved solute in a solvent. The droplets undergo rapid evaporation of solvent upon contact with turbulent hot gas flow, favouring the formation of amorphous particles. The process is well suited for production of amorphous solid dispersions. However, amorphous solid dispersions can be unstable and the API can transform to crystalline form when exposed to high temperature and humid conditions, which reduces its efficacy. Furthermore, the size distribution of particles, density and moisture content also effect the product performance. By optimising the input operating conditions in the spray dryer and feed composition, it is possible to control the dried powder attributes.

Modelling can be used to assess the impact of feed composition and dryer operating conditions on the size distribution, density, moisture and crystallinity of spray dried particles in a cost effective manner. In this study, modelling of a lab-scale co-current spray drying tower utilising two-fluid nozzle for feed atomisation is carried out to predict dried powder characteristics. The turbulent air flow is modelled using the Computational Fluid Dynamics (CFD) approach and coupled with a detailed droplet drying model to predict drying rate as well as concentration gradients within the droplet/particle. A distributed modelling approach is applied to predict amorphous to crystalline transformation within the particles undergoing drying, utilising the glass transition temperature. The drying and crystalline transformation model is applied to feed containing lactose-water solution and the results are validated with data available in literature. The modelling approach can be applied to an industrial scale tower to predict optimised operating conditions.

508577 Quality By Control of Dry Granulation Process for Continuous Tablet Manufacturing

Abstract:

Integrated continuous processing is currently a key focus area in downstream oral solid dose manufacturing [1]. Quality by Control involves process understanding and real-time monitoring of stream flow rates and critical quality attributes from the inlet feed to the product, including the intermediates to maintain the production throughput and control the product quality standards of a continuous process.

Dry granulation is one of the intermediate unit operations added for improving material handling and blend uniformity. Since the particle size distribution and bulk density of the resulting granules affect the operation of the tableting process, Controlling the relative density of ribbons is essential for the desired properties of the granules. In this study, a process model of the roller compactor accounting for the compaction as well as the flow rate through the system is used to predict the ribbon density to enable model based control. The model combines the dynamics of material flow from Hsu et al [2] and Reynolds et al [3] and solid mechanics suggested by Liu and Wassgren [4]. In-line measurement of the ribbon density using a microwave sensor [5] enables the validation of the model and for controlling ribbon density at the set flow rate. The particle size distribution, blend uniformity of the granules, along with the mass flow rate of the materials entering the tablet press are measured using appropriate inline sensors.

Blends comprising of acetaminophen and microcrystalline cellulose at inlet flow rates of 8-12 kg/h and varying roll pressures are used for real-time monitoring of ribbon density. The process variables, ribbon density and mass flow rate are recorded and monitored using DeltaV DCS.

This work enables demonstration of the implementation of in-line ribbon density control, a key intermediate quality attribute for a dry granulation integrated continuous tableting line. The integrated process model along with real time monitoring of granule size distribution and the mass flow rate enable the development of real-time material tracking and control strategies for the integrated continuous tableting line.

References:

1. https://blogs.fda.gov/fdavoice/index.php/2016/04/continuous-manufacturing-has-a-strong-impact-on-drug-quality/
2. Liu Y, Wassgren C. Modifications to Johansonâ€™s roll compaction model for improved relative density predictions. *Powder Technol*. 2016; 297:294-302.
3. Reynolds G, Ingale R, Roberts R, Kothari S, Gururajan B. Practical application of roller compaction process modeling. *Comput Chem Eng*. 2010;34(7):1049-1057
4. Hsu S-H, Reklaitis G V., Venkatasubramanian V. Modeling and Control of Roller Compaction for Pharmaceutical Manufacturing. Part I & II: Process Dynamics and Control Framework. *J Pharm Innov*. 2010;5(1-2):14-36
5. Gupta A, Austin J, Davis S, Harris M, Reklaitis G. A Novel Microwave Sensor for Real-Time Online Monitoring of Roll Compacts of Pharmaceutical Powders Online-A Comparative Case Study with NIR. *J Pharm Sci*. 2015;104(5):1787-1794.

508507 Twin Screw Granulation: A Step Towards Continuous Processing

Abstract:

TWIN SCREW GRANULATION: A STEP TOWARDS CONTINUOUS PROCESSING

Rachael Shinebaum[a], Andrew Ingram[a], Hannah Batchelor[a], Gavin K. Reynolds[b], Ian Gabbott[b]

[a.] School of Chemical Engineering, University of Birmingham, Edgbaston, B15 2TT

[b.] Pharmaceutical Technology & Development, Astra Zeneca, Macclesfield, SK10 2NA

Within the pharmaceutical industry, a shift towards continuous processing to produce granules, away from traditional batch-wise techniques is underway. Twin Screw Granulation (TSG) is a continuous process which, in recent years, has been the subject of much research to increase knowledge in its role within solid oral dosage form manufacturing. TSG reduces plant footprint, reduces process development costs and allows ease of scale-up to full production. Within the TSG process there are a large number of variable process parameters including: screw element configuration, screw speed, liquid feed quantity and powder feed rate. Further work is required to gain a deep understanding of the mechanisms and process factors required to predict the characteristics and quality of the resulting granules required by the FDA regulations, for example Quality by Design (QbD) and Process Analytical Technology (PAT).

This study investigates the impacts of liquid to solid ratio and screw configuration on the properties of the granules produced by the twin-screw granulator and their subsequent compaction behaviour and tablet properties. A blend of lactose, microcrystalline cellulose (MCC) and hydroxypropyl cellulose (HPC) was granulated using six different liquid to solid ratios and five different screw configurations. The granules were characterised with X-Ray micro tomography to measure porosity and structure, while shape and size analysis was carried out via QICPIC particle size analyser. Granule size increased dramatically with increasing liquid content and kneading zones within the barrel. Tabletability was shown also to be affected by the granule manufacturing conditions and was correlated with the granule.

506457 A Novel Metrology for Polymodal Particle Size Distributions and Its Beyond Application in Twin Screw Granulation

Abstract:

Patient preference is for the oral administration of drugs as tablets, which also provides simply control over the dosage given. However, in the majority of the cases, the drug cannot be tabletted directly due to inappropriate powder properties and granulation to control powder properties is one of the key steps in a pharmaceutical tablet manufacutirng process. This is further complicated by the addition of excipients to modify and control physical properties but which also can influence drug release in the patient.

The most common granulation approach is the wet method which is traditionally done as batch unit operation. Currently, there is a drive within the pharmaceutical industry to adopt continuous processes and one of the most commonly used pieces of equipment for this purpose is the twin screw granulator (TSGs). TSGs provide shorter residence times, greater flexibility in granule properties and the ability to vary the required throughput without associated hazards (1).

The equipment can be described as a barrel which contains two co-rotating screws which act as agitating conveyor belts. Initially the powder is fed into the barrel, before the liquid binder is added to induce granulation. The co-rotating screws can be easily modified by configuring different screw elements or changing their position along the barrel length. Different screw configurations will modify how contact between the particles is produced and, therefore, the properties of the final granules. A broad range of outcomes can be achieved by altering these elements as well as other factors such as liquid/solid ratio, feed rate and rotation speed.

Amongst all the possible granule properties particle size is critical in pharmaceutical manufacture and currently this is studied by separating the sample into discrete size classes and measuring the relative amount of particles within a specific size interval which is expressed as particle size distribution (PSD). Unlike other properties such as friability or flowability, particle size distributions are expressed as a curve, which does not allow its direct examination as a quality attribute or a process control variable.

The most common solution to compare volume particle size distributions is the use of the d50 combined with the span calculated by gathering d10, d50 and d90 which are the intercepts for 10, 50 and 90% of the cumulative volume. The closer the span value is to zero the more homogeneous the particle size distribution. However, this analysis is only acceptable when the particle size distribution is unimodal and therefore its shape is lognormal, otherwise it will introduce a considerable error when the distributions show more than one peak and the peaks are located around the mean diameter.

Multiple peaks distributions can be very common in research before the equipment has been optimised. However, they are very difficult to compare without a numerical factor. As a solution, a methodology based in the hyperbolic tangential method (2) was developed and used for measuring homogeneity of particle size distributions. The method is able to calculate the homogeneity of PSDs with different shapes as a percentage allowing easy numerical comparison. On the contrary to span, the method responds to different modifications such as the addition of peaks, increments on the variation of the distribution or discrepancies between the main diameter and main particle size class. Using homogeneity as a quantified quality attribute leads to a better understanding of powder technology and its possible implementation as characterisation tool in the design and control of wet granulation systems.

The improvement of this method with respect to the traditional measures such as span was demonstrated through the comparison of PSD curves with different shape but similar span (3). In addition, the potential of the quantification of homogeneity was demonstrated through the application to simple wet granulation with two different excipients. In both cases, it was demonstrated that knowledge of the d50 alone does not give enough information for the ideal conditions to operate or which operational parameters have more influence on the process.

Understanding of the underlying process mechanisms of a new technology requires to be accomplished with a systematic scientific risk-based methodology. In the pharmaceutical industry, this approach is recognised as Quality by Design (QbD) and it has been recommended for adoption (4). Hence, using homogeneity as characterisation tool for PSD can be potentially advantageous. It could be used in the comparison of different batches and technologies on both a research and industrial scale as well as being a useful complement to other granule properties such as flowability or strength in the optimisation of tabletting and associated processes.

In a research context, it can be employed by defining a desired diameter operating point and controlling the homogeneity around that point. That will allow the identification of when the process is within product specifications. The benefit of this parameter has been applied to find the most homogeneous points for three different diameters (400, 600 and 800 Î¼m) of Î ±-lactose monohydrate and study the differences in the granular structure.

Studying the internal parameters of the granules will help to understand the formation of the granules within TSG and to compare performance with conventional batch equipment. The microporosity of the samples is determined by Quantrome Autosorb IQ2 and the granules are scanned by MULTISCALE X-ray nanotomograph SKYSCAN 2211 and processed in Matlab and Statistics Toolbox.

Despite the various advantages of twin screw granulation over batch granulation, this technology is not yet fully applicable outside the research and development phase. As a result of this, it has not been fully adopted in large scale industrial pharmaceutical manufacture. As such the current study aims to increase the understanding of this technique, it is hoped the study of the granular internal structure and the development of a homogeneity factor will enable the development of twin screw granulation as a suitable technique for widespread use both in academia and industry.

REFERENCES:

(1) Iveson, Simon M., et al. "Nucleation, growth and breakage phenomena in agitated wet granulation processes: a review." Powder technology 117.1 (2001): 3-39.

(2) Passe, T.O.R.E. "Grain size distribution expressed as tanh-functions." Sedimentology 44 (1997): 1011-1014.

(3) Torrecillas, Carlota Mendez, Gavin W. Halbert, and Dimitrios A. Lamprou. "A novel methodology to study polymodal particle size distributions produced during continuous wet granulation." International journal of pharmaceutics519.1 (2017): 230-239.

(4) ICH Expert Working Group. "ICH Harmonized tripartite guideline. Pharmaceutical development Q8 (R2)." International Conference on Harmonization of Technical Requirements for Registration of Pharmaceuticals for Human Use. 2009.

AUTHOR INDEX

AUTHOR INDEX

AUTHOR INDEX

AUTHOR INDEX

AUTHOR INDEX